Webb Telescope's Revelations:

Is Humanity Ready for the Truth?

TD McRoy

ISBN: 979-8-9890620-0-3   *(Hardcover)*

ISBN: 979-8-9890620-2-10   *(Paperback)*

Year of Publication: 10/22/23 *(Official release date)*

Credits: NASA, ESA, CSA, and STScI

Cover Design: by TD McRoy

Illustrations: by TD McRoy -Iroquois Technology -Dynamic Filtration

Dedication

To David Manning –an early mentor who greatly influenced my journey and instilled in me a deep sense of self-confidence. David who was the father of a childhood friend invited me to speak at his rotatory club during my most informative years. That invitation was truly profound and established a true since of not only who I was, but the potential to who I could be. For that I am forever in David Manning's Debt.

To Peggy J. Collins,

In the tapestry of life, some threads shine brighter and bolder than the rest. You, dear Peggy, are one of those resplendent threads in the weave of my existence. From a fragile age, when the world seemed vast and overwhelming, you wrapped me in the comforting embrace of a family.

To call you a mentor feels inadequate. To me, you are the second mother my heart found, a beacon in the stormy nights of my early adult life. At times when I grappled with the weight of independence and the pressures of making ends meet, you ensured that I was never truly alone by opening up your home to me. In an era when it was rare for someone to embrace those outside their immediate kin, you defied conventions. You saw a young man, vulnerable and striving, and you opened not just your home but also your heart. Thank you for being the inspiring woman you are.

To Jimmy Taylor,

In the winding roads of the 1990s, as I embarked on my journey of owning a trucking company, it was your enduring passion for Peterbilt trucks that became my beacon. Your unwavering dedication, not swayed by societal expectations, taught me the profound value of following one's heart. While many drift in the quest for career satisfaction, you, Jimmy, had the rare gift of recognizing your calling early in life. Your love affair with trucks, which began in your youth, remains as fervent today as it was fifty years ago. As I dedicate this book, I am reminded of your zest and zeal, which has been an inspiration. Many of us search our entire lives for that one thing that sparks our souls. You, having found it so young and cherishing it undiminished over decades, are a testament to true passion.

"To you, and to my dear sister Gloria Taylor, who stands by your side in every journey you undertake, this book is dedicated. As are the two cherished counterparts, David and Peggy."

May we all find our passion, as you did, and hold onto it for a lifetime.

Acknowledgments

I would like to extend my gratitude to all researchers from around the globe who have supported and closely followed our research. A heartfelt thank you to the members and fans during our formative years of QEIN research.

Foreword

This non-fiction book presents detailed figures extracted from two captivating photographs taken by the James Webb Space Telescope. Leveraging the advanced Iroquois technology, these profound images of QEIN were disclosed and later enhanced by the state-of-the-art Dynamic Filtration — a program intricately developed to align with the Iroquois Technology. Although the images may evoke notions of extraterrestrial origins, they offer a mere inkling of a potential alternate reality. The author earnestly underscores that, despite the rigor of his research institution's methodologies, there exists no definitive evidence of such alternate existence within our understanding. While these images were deciphered using scientific methods, the author emphasizes that this publication's primary intent is entertainment. Any assertions about the figures remain conjectural and are neither validated by NASA nor the researchers of the James Webb Space Telescope. Such interpretations are the exclusive purview of the Dimensional Anthropology Museum.

Certifications and Education

- Autism and Neurodiversity, Curtin University

- Natural Disasters, McGill University

- Astrophysics: The Violent Universe, ANU, Australian National University *(The best course ever)*

- Strategic Decisions for Project Leaders, University of Maryland

- Humanitarian Response to Conflict and Disaster, Harvard University

- Meteorology: The Science of Weather, Harvard University *(The second best course ever)*

Epilogue

Epilogue: A Journey from Uncertainty to Discovery

When I embarked on the journey to create the Iroquois Technology program, doubt overshadowed my expectations. The idea of detecting concealed particles within film footage felt like a page out of a science fiction novel rather than a chapter from a scientific journal.

Yet, as fate would have it, our first monumental discovery in 2007 unveiled an image of a native figure seemingly from the Iroquois nation. Despite the figure's obscurity, his attire spoke volumes. This seminal moment inspired the name "Iroquois Technology".

The program's capabilities were put to the test in films like "A Haunting on Finn Road" and then again in "The Devil's Grove: A Haunting on Brockway Street." During the latter, the director provided us with film clips from a house alleged to be active with unnatural events. Without prior insights into the site or its stories, our program astoundingly pinpointed the images of a man and a young boy in an upstairs bedroom. This revelation not only astonished the director but also reshuffled our understanding of the program's capabilities. The realization that Iroquois Technology could detect images in real-time catapulted our Dynamic Filtration program's progress by two years.

Our confidence surged when the program accurately identified figures of Asian descent from a nondescript forested film clip, later learned to be sourced from China.

As we share our findings and experiences with the world, we pledge a significant portion of this book's proceeds to The Bertha Project, dedicated to extending a helping hand to those affected by natural disasters. Our vision doesn't stop there.

We're collecting donations to build a unique online museum complete with an IMAX theater to display the astounding 100 images unearthed by Iroquois Technology. For enthusiasts eager to bring a piece of this revolutionary tech into their homes, we offer exclusive posters at the Iroquois Technology Store on Etsy.

This journey has been one of unexpected turns, revelations, and commitment to a cause greater than oneself. As we look to the future, we will have to do it without Iroquois Technology. This publication serves at the final chapter in Iroquois Technology.

Glossary

- **Iroquois Technology**: Is an AI program that was created by TD McRoy back in 2007. The program is thought to detect the signature of particles hidden within film footage. The particles were believed to be muon and tau neutrinos, but this could never been confirmed. We we could confirm that the program detected odd spaces of particles in some films where unidentified activity was detected.

- **Dynamic Filtration**: This program was created in 2018 to help enhance the images that Iroquois Technology discovered. Simply put, the program made the dark in an image darker and the whites in an image brighter. Often times the eyes of a figure seem a tad bit brighter than in other areas of the discovery. We could never figure out what they were, but it did help us detect the human and animal-like figures seen in many film clips.

Bibliography

Greene, Brian. *The Elegant Universe*.

Hawking, Stephen. *A Brief History of Time*.

Homer. *The Iliad*.

Homer. *The Odyssey*.

NASA. "James Webb Space Telescope*.

Sagan, Carl. *Cosmos*.

Tyson, Neil deGrasse. *Astrophysics for People in a Hurry*.

Tyson, Neil deGrasse, *Welcome to the Universe*. Princeton University Press.

Index

About the Author

"TD McRoy is a versatile professional with a deep-rooted love for science. For over two decades, he dedicated himself to a firefighting career. Simultaneously, unbeknownst to many, he pursued his academic passion for physics. A pivotal, unexplained event in 2005 steered him towards his present domain of research. This mysterious episode is detailed in his book, "33 Miles from Washington," where he recounts strange occurrences that eventually paved the way for the inception of Iroquois Technology."

Other Works

The book "33 Miles from Washington" can be purchased as a paperback edition on Amazon. We're actively fundraising for the forthcoming Dimensional Anthropology Museum, set to launch online. Once operational, guests will immerse themselves in a 3D virtual environment, navigating the museum's corridors and joining others in a virtual IMAX theater.

This experience aims to recreate the sensation of sharing a theater with loved ones, regardless of their global location. Our initiative seeks to transform cinematic experiences, especially catering to individuals with mobility challenges, those on the autism spectrum who might find traditional theaters overwhelming, or families and friends separated by distance. Our aspiration is for our IMAX to be a pioneering platform, eventually hosting mainstream Hollywood releases for those who, due to medical constraints, are unable to attend traditional movie screenings.

Introduction to Bold Emphasis Formatting (BEF):.

Introduction to Bold Emphasis Formatting (BEF): In this book, we introduce a unique approach to content presentation known as Bold Emphasis Formatting (BEF), developed by the Dimensional Anthropology Museum. While traditional citation styles offer structured frameworks for academic writing, BEF seeks to enhance reader engagement through a visually distinctive and expressive presentation.

By incorporating bold text for emphasis and structuring sections with a personalized touch, BEF aims to create a more immersive reading experience. Reasons for Creating Bold Emphasis Formatting:

1.

Enhanced Emphasis: Traditional citation styles often rely on a uniform approach, which can sometimes limit the ability to emphasize key points effectively. BEF capitalizes on the power of visual elements by utilizing bold text strategically to highlight essential concepts, enabling readers to swiftly grasp the core ideas within the content.

2.

Reader Engagement: As the creators of the Dimensional Anthropology Museum, we understand the significance of captivating narratives and immersive experiences. BEF seeks to mirror the immersive experience of our exhibits by fostering a connection between readers and content. By presenting information in a unique manner, BEF stimulates curiosity and encourages readers to explore further.

3.

Personalized Visual Structuring: BEF brings a personalized touch to the formatting landscape. Unlike standardized citation styles, BEF allows authors to adapt formatting to their content's specific needs. Through section structuring that aligns with the flow of ideas, BEF assists readers in navigating content effortlessly, fostering comprehension and retention.

4.

Innovation and Distinction: Just as the Dimensional Anthropology Museum pioneers innovative approaches to understanding our world, BEF reflects our commitment to pushing boundaries in the realm of content presentation. This distinctive formatting style sets your reading experience apart, showcasing your individuality as an author while maintaining clarity and coherence. As you embark on this reading journey, embrace the visual storytelling of Bold Emphasis Formatting, crafted to inspire curiosity, enhance comprehension, and offer a fresh perspective on the art of conveying ideas.

Contents

CHAPTER ONE

Prior to perusing the contents of this publication, readers should be apprised that some images may be perceived as exhibiting spiritual characteristics. We affirm that no augmentations have been made to the figures visible in the James Webb Space Telescope photographs, as analyzed by Iroquois Technology. Each representation can be cross-referenced with its original space photograph .

The implications of these images remain ambiguous and may perpetually elude definitive interpretation. However, we can assert that these phenomena were not anthropologically fabricated but are a product of cosmic processes discovered in 2023.

In this research paper, we introduce quantum figures designed primarily for illustrative purposes. The algorithms employed by Iroquois Technology to detect celestial entities have not been subjected to comprehensive scientific scrutiny, nor have they been acknowledged as established scientific tools. This perspective is echoed by the Dimensional Anthropology Museum. Our team perceives the software more as illustrative tools than as rigorous scientific instruments, akin to the functional props used in the entertainment industry. While Iroquois Technology shows promise, it lacks formal scientific endorsement. Our decision to abstain from further validation was driven by the desire to maintain autonomy and ensure ethical management of its profound implications. We encourage readers to form their own interpretations, and our organization respects all viewpoints.

The James Webb Space Telescope embarked on its monumental journey on Christmas Day, 2021, marking a new era in space exploration.

This mighty instrument, the largest of its kind launched into the depths of space, seeks to illuminate the earliest moments of the Universe, tracing back to the first spark of galaxies and stars following the Big Bang. Its recent captures have unveiled an astonishing array of cosmic snapshots, offering a glimpse into the genesis of everything.

Among its early comprehensive explorations of the cosmos, the Telescope accumulated 250 hours of exposure, greatly broadening our understanding of the Universe. These images presented a variety of galaxies, some even exceeding the size of our own Milky Way, proving space to be a far more crowded neighborhood than once presumed. The gravitational dance of galaxies, as they interact and merge, came to life in these unprecedented images. A digital tour through some 130,000 previously unseen galaxies was facilitated by an astrophysicist based at the University of California, Santa Cruz.

Prominent among these celestial bodies was a galaxy bearing a coincidental resemblance to a rose, earning it the nickname of the 'cosmic rose'. Several other galaxies were depicted in the throes of

gravitational disruption, signifying their merger and highlighting the dynamic nature of the cosmos.

The James Webb Space Telescope embarked on an ambitious mission named the Advanced Deep Extragalactic Survey (JADES), leading to the discovery of the most distant galaxy currently known, lying more than 33 billion light-years away. This early universe inhabitant, estimated to have formed a mere 320 million years post-Big Bang, existed when the Universe was only 2% of its current age. Despite its significantly lesser mass compared to the Milky Way, it possesses a high star-formation rate.

The preparation and launch of this $10 billion observatory caused considerable anticipation in 2021.

The telescope, named after an early NASA Administrator, features a large sun shield and a 21-foot mirror of gold-plated segments that work in unison to gather light and form a single, comprehensive image.

This awe-inspiring telescope embarked on its mission, transcending from the verdant tropical rain forests to the very edge of time, casting its gaze on galaxies everywhere, challenging the notion of an 'empty' sky. Despite initial assumptions categorizing some detected points as artifacts or strange stars, it was revealed that these were in fact numerous, previously unobserved galaxies.

This revelation brings to the fore the vast unknowns of the Universe, including the elusive dark energy and dark matter, unseen forces making up about 95% of the Universe. Despite such profound ignorance, the field of astronomy tirelessly ventures on, unveiling previously unimaginable details of the cosmos.

Supernovae, the remnants of exploded stars, provide the raw materials vital for life – the iron in our blood, the calcium in our bones, and the oxygen we breathe. The telescope's ability to reveal unprecedented details at the core of these explosions significantly adds to this understanding.

Furthermore, the telescope is adept at detecting infrared light, the heat radiation from the dawn of time. This light, invisible to the human eye, is transformed into a colorful spectacle by astronomers and science imagers at the Space Telescope Science Institute, unveiling cosmic structures such as star clusters and nebulae.

The contributions of the James Webb Space Telescope have already been significant. It has detected carbon dioxide in the atmosphere of a distant planet and potentially identified galaxies that challenge current theories of early Universe formation.

An astrophysicist from the University of Colorado Boulder proposed that five giant galaxies appear to have formed much too quickly after the Big Bang, a potential discovery that could lead to a revision in the timeline of galaxy formation.

In summary, the James Webb Space Telescope's extraordinary capabilities hold the promise of unveiling every galaxy ever formed in the Universe, signaling a transformative shift in our comprehension of the cosmos. This remarkable instrument's launch may even delineate two distinct epochs in the history of astronomy: before and after its deployment.

However, the Dimensional Anthropology Museum aims to add another epoch to this progression, representing the revelation of images presented in this publication. These images, captured by Iroquois Technology, are poised to contribute significantly to a better understanding of the Universe, marking a pivotal decision in the field of astronomy as to what affects devices like the James Webb Space Telescope may have on energy unknown. This observatory could operate for up to 25 years, potentially long enough to revolutionize our understanding of space, time, and the origins of life.

Beyond the observable, the Universe holds remarkable revelations yet to be unveiled. The universe weaves intricate connections between all its constituents. This chapter uses Iroquois Technology to comprehend celestial phenomena beyond the electromagnetic spectrum. The publication emphasizes correlations between humanity and stars, highlighting shared elements and interconnectedness.

Both humans and stars share origins in ancient stars, as every atom in our bodies was forged through nuclear fusion and scattered across the Universe during the death of stars.

Energy Production: Humans and stars are energy producers. Stars generate energy through nuclear fusion, while humans produce energy through cellular respiration.

Stars and humans follow similar life-cycles, starting from birth, going through various stages of existence, and ultimately ending. Both stars and humans are intricate systems with layers of varying properties and compositions.

The fundamental similarities between stars and humans underscore the interconnected nature of the Universe. We are not merely inhabitants but an integral part of it, sharing a cosmic lineage with the stars. The distinctions we draw between 'us' and 'them' are blurred in light of this understanding.

This manuscript's primary ambition is not merely to reiterate the discoveries made by the James Webb Space Telescope. Instead, it seeks to illuminate the unique findings extracted by deploying Iroquois Technology on the images procured from the Webb Telescope. As it stands, this technology has been applied to just one photograph from the Webb Telescope, but the outcomes have been striking. These remarkable findings lend credence to the possibility of the authenticity of Greek mythology, opening new and compelling areas for future investigation.

Scholars at the Dimensional Anthropology Museum propose that the Webb Telescope's discoveries within the 95% of dark matter around galaxies may testify to the continual existence of humankind. Accepting thermodynamics, it is argued that energy cannot be created or destroyed within a closed system, implying that human energy transitions to another form or location upon death. The James Webb Telescope's findings could potentially represent a celestial repository of these energy transformations.

This manuscript aims to illuminate unique findings derived from Iroquois Technology applied to images procured from the Webb Telescope. The technology has been used on a single photograph, yielding striking outcomes.

These findings provide credence to the authenticity of Greek mythology and open new areas for future investigation.

Before revealing the images discovered by Iroquois Technology, let me provide some background. The technology was developed by researcher TD McRoy in 2007, hypothesizing the existence of hidden muon and tau particles, or energy signatures, within film footage. By detecting the particles' distinct pattern within film footage, Iroquois Technology overcomes the need for a supercollider for detection.

The crux of Iroquois technology lies in the characteristic pattern of muon and tau particles, which manifest covertly within film footage. Similar to nuclear detonations displaying a specific morphology in mushroom clouds, the quest for these particles within film footage is guided by identifying their unique pattern, not observed in all films.

Utilizing a refined approach to particle physics, we wish to disseminate our observations to the world. An intriguing capture from the James Webb Space Telescope displays enigmatic representations reminiscent of early human likenesses in a galaxy aged approximately 4.6 billion years, defying its age. Notably, one image bears resemblance to a deity portrayed in the cinematic representation. Drawing parallels from historical literature, the astronomical capture might suggest entities like Thor, Zeus, or possibly Odin, originating from the James Webb Space Telescope's images.

1. In this intriguing photograph below, we are presented with a prominent figure resembling a character from Norse mythology. While it is crucial to acknowledge that our understanding is limited to written descriptions of ancient deities, this depiction bears resemblance to Odin, the renowned All-Father. As a significant chief deity, Odin held authority as the ruler of Asgard, the divine realm, and epitomized attributes of wisdom, knowledge, and martial prowess.

His countenance exudes wisdom, characterized by a bearded visage, projecting a multifaceted and enigmatic persona that elicits curiosity and captivation.

Figure 1: Odin

CHAPTER TWO

2. In the subsequent image sourced from the same celestial region, an entity resembling a member of the Canidae family is evident. The elongated rostral anatomy characteristic of a canine is discernible to the right of the anthropomorphic visage with facial hair. This instance is not the first time an image with features akin to a long canine snout has been identified within this constellation. There are other long snout dogs that contribute to the celestial collage.

In Greek mythology, the gods are not typically depicted as having pet long-snouted dogs. There are no prominent myths or stories that specifically mention Greek gods owning or having long-snouted dogs as pets. The Greek gods are usually associated with various animals and creatures, but long-snouted dogs are not among them.

Some significant animals associated with Greek gods include:

1. Zeus – Eagle: The eagle was considered Zeus' sacred bird, often depicted as his companion or as a symbol of his power.

2. Hera – Peacock: The peacock was associated with Hera, the queen of the gods, symbolizing her regal and majestic nature.

3. Poseidon – Horses: Poseidon, the god of the sea, was closely associated with horses, often depicted riding a chariot pulled by majestic sea horses.

4. Artemis – Deer: Artemis, the goddess of the hunt, was often shown accompanied by deer, as they were sacred to her.

5. Dionysus – Panthers and Tigers: Dionysus, the god of wine and revelry, was sometimes depicted with panthers or tigers, symbolizing his wild and unpredictable nature.

While long-snouted dogs are not specifically linked to any Greek gods in mythology, it is essential to note that ancient myths and artistic representations of deities varied across different regions and time periods. Therefore, there might be obscure or localized references to long-snouted dogs in some lesser-known myths, but they are not widespread or significant in the overall depiction of Greek gods in classical mythology.

One of the wonders of Iroquois Technology is the visual truth it provides about the past. If one is to believe that the images discovered by the program are authentic depictions of ancestors, it becomes clear that the image is of a man and his trusted dog standing beside him.

Figure 2: Odin

The subsequent representation is notably compelling, appearing to depict an indigenous individual adorned with a cranial accessory reminiscent of those portrayed in cinematic interpretations of Hellenistic combatants. The clarity of this image is somewhat inferior when compared to the aforementioned anthropomorphic and canine figures. However, with meticulous observation and analysis, one can discern the semblance of a melanin-rich Hellenistic warrior.

Our researchers have employed a proprietary program, termed Dynamic Filtration, which enables the revelation of hidden particles within images discovered by Iroquois Technology that would otherwise remain unseen by the naked eye.

The first piece in this triptych displays the genuine visual data as recorded by Iroquois Technology. The subsequent image accentuates the cranial decoration indicative of the Hellenistic combatant. The final portrayal in the set is a digital refinement, offering detailed contours and nuances of the individual's facial structure, showcased in the green section. Pay particular attention to the combatant's expansive eye located in the lower segment of the green depiction—it's reminiscent of the watchful eye of a hunter from his native land.

Figure 3: Hellenistic combatant

In the collection, the fourth image on the next page is particularly striking. Identified by the program, its pale, almost alabaster-like face stands out more clearly than the others in the series. This male figure, distinguished by a unique triangular headpiece, suggests a position of leadership or symbolic importance. His face, reminiscent of Omanus, contrasts sharply against the surrounding darkness. A notable feature is what appears to be a large rectangular earring hanging from his left ear, accompanied by a bone accessory near the shimmering jewelry. Subtly, in the photo's upper right corner, another face seems to gaze towards the James Webb Space Telescope, hinting at a conversation about the unexpected presence.

Figure 4: Triangular Cranial Being

The fifth image in this compilation hosts an array of captivating features, warranting a detailed sequential examination. Initially, the gray arrow, located towards the photograph's lower region, delineates what seems to be the ruby-red lips of a feminine figure. The subsequent arrow above reveals the commencement of her nasal structure. Intriguingly, researchers noted the manifestation of what appears to be Acne vulgaris on the figure's visage.

Acne vulgaris ranks within the top three dermatological conditions encountered globally, with historical documentation and treatment methods tracing back to ancient Greek and Egyptian periods. Given the previous identification of a figure apparently donning a Hellenistic helmet and another bearing resemblance to Odin, one might hypothesize that this female figure could be of Greek or Egyptian origin.

Figure 5: Ancient Greek or Egyptian Woman

The figure, framed by the yellow rectangle on the next page, is deeply captivating. It portrays the face of a man, perhaps superimposed by another similar visage, suggesting a form of rebirth or renewal.

This individual has long, dark hair and is adorned with a V-shaped garment, reminiscent of a chiton. Strikingly, there are marks on the garment that hint at wounds, with distinct crimson traces. The representation compels one to ponder the historical or symbolic significance of this man.

The perception of this image might not be immediately evident due to its complex composition. Repeated examination of the photograph is recommended to discern the specific elements referenced. It's not uncommon for individuals to not recognize intricate details during an initial viewing. The figures within the image exhibit multi-dimensional layers, and human vision isn't always primed to differentiate such intricacies upon first glance. To assist readers in identifying landmarks on the image of the man, start by pinpointing the 'V' shape in the garment. From there, move vertically upwards to find the neck, leading you directly to the head. Surrounding the 'V'—to its left, right, and directly below—you will notice regions marked in red, indicating possible signs of distress or harm.

To the right of that rectangle is another one that presents a side view of what seems to be a mother partridge. It is intriguing to note that this avian figure, situated next to the man in the tunic, represents the sole depiction of a bird in the image. The large eyeball located at the upper left is creepy.

Figure 6: The Chiton Man

The next visual representation in this compendium we tentatively designate as "Pegasus." Regardless of whether this truly represents the famed alabaster steed from Hellenistic lore, the close proximity of other Greek-related images prompts us to hypothesize, albeit speculatively, that this recently identified galaxy could have been visible to ancient Greeks, potentially serving as a celestial canvas for their divine entities during a prior cosmic epoch.

This galaxy, at some point, may have been as close and observable as our current lunar body.

If this holds true, it could shed light on our recent observations. Considering that early Hellenistic individuals might have gazed upon the nocturnal sky, it is plausible for them to have perceived what Iroquois Technology is currently identifying. It is essential to note that astrophysicists have claimed that galaxies are defying the laws of gravity and rapidly moving away from each other. This suggests that this particular galaxy may have been aligned with Greece back when the Greeks were attributing claims to these Gods.

Figure 7: Pegasus

The ensuing photograph constitutes a portion of the third image in the sequence, highlighting the Hellenistic combatant. A feminine figure, assimilated within the warrior's structure, has been singled out for a more comprehensive analysis. If this female depiction is authentic, it could signify her remarkable standing, considering her singular existence amidst a galaxy predominantly occupied by male figures.

"On the following page, you'll find the authentic photograph taken by the JWST. As the author, I purposefully delayed unveiling the precise location where the Iroquois Technology program registered its initial hit. Out of all possible human-like figures the program could have identified, it's striking that the first was one resembling Odin—a figure many might envision as the ruler of such a realm, should it exist. This compelling image is but a taste of the enigmas enclosed within these pages. I chose to reveal its true significance later, hoping to spark a burning curiosity in you about the proof behind the cover image, enticing you to keep searching for an answer that must surely lie ahead.

The figures we uncovered early on set the stage. They invite readers to immerse themselves, to grapple with potential realities, and to contemplate the significance of such mystifying figures. By the time you, dear reader, confront the narrative implications of this Odin-like figure, you're poised to comprehend not just his appearance but his profound essence—and to ponder the tantalizing possibility of this tale being rooted in truth. I remain convinced that there's a genuine story here, brimming with wonder. The magic of this tale doesn't just reside in its events, but in the very act of unveiling brought forth by the JWST.

To the left is a cropped close-up of the galaxy, with our 'Odin' figure adjusted for clearer viewing. The larger image on the right is the original JWST capture. We've highlighted the location of what we've come to call Odin. Every detail you see on the cover is extracted from this original and can be verified. The intricacies of the 'Odin' figure were further illuminated by a complementary program, Dynamic Filtration. All details on the cover are faithfully represented within the original and stand ready for cross-referencing."

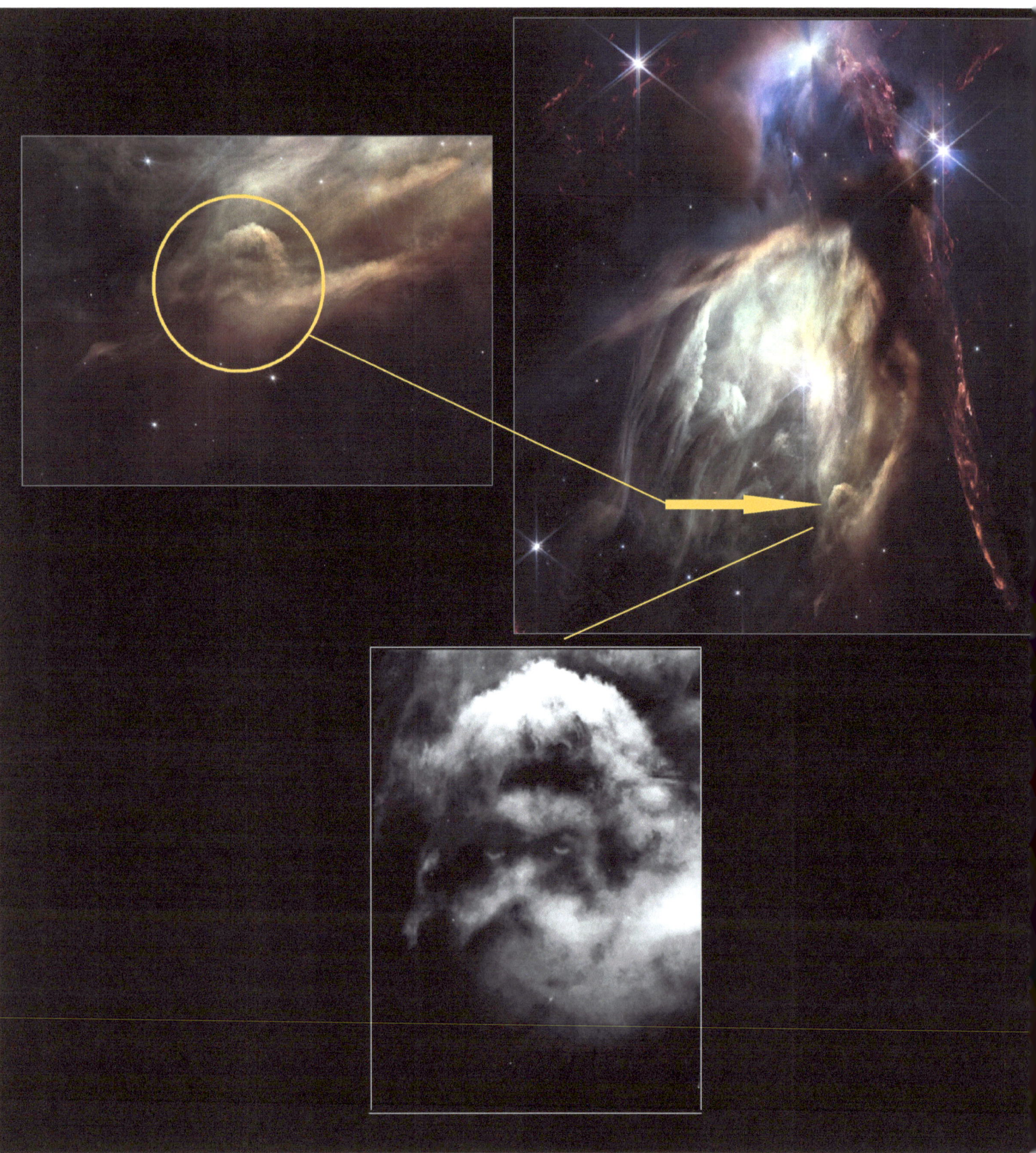

Figure 7a: Original Odin-NASA

CHAPTER THREE

Characterized by her long, dark hair, one might surmise that this influential woman commanded respect among her contemporaries. The historical figure Lady Godiva springs to mind in this context. Known to have lived between 1066 and 1086, she was an Anglo-Saxon noblewoman, married to Leofric, Earl of Mercia.

Today, she is primarily remembered for a legend that dates back to at least the 13th century. According to this tale, she rode naked—concealed solely by her long hair—through the streets of Coventry to secure a reduction of the onerous taxation that her husband, Leofric, levied on his tenants.

Figure 8: Lady Godiva

The pair of images presented on the next two pages depict a lateral perspective of ancient military personnel equipped with a high-lid helmet, complete with a chinstrap. These types of pit or sun helmets were worn by imperialist nations during the so-called "scramble for Africa" and the European race for colonies across the world. This pattern was worn by men in red coats as they faced off against the Zulu in South Africa, the Italians in Ethiopia, and the French in Indo-China.

The visual clarity of this figure in the photograph is not immediately apparent. However, we advocate for viewers to engage in repeated examinations of this image over time, which could facilitate increased perceptual clarity. The second photograph is accompanied by an overlaid chart and a numerical system designed to aid the viewer in the identification of various components of the soldier.

Upon examining the photograph, particularly at figure number six, it is noticeable that there appears to be a prominently displayed left hand. Both the pinky and its adjacent finger are adorned with what seem to be fingertip rings. It's worth noting that the extensive adornment of fingers with jewelry has been a practice in various cultures. Such embellishments could encompass the usage of rings on diverse segments of the fingers. While not prevalent in contemporary society, it is plausible that ancient cultures may have harbored a particular emphasis on fingertip decorations across their entirety.

Figure 9: The Imperial Man of Nations

23

Figure 10: The Imperial Man of Nations

The photographic compendium contained within this publication emanates from a singular galaxy formation imaged by the James Webb Space Telescope. However, capturing clusters of stars does not unequivocally elucidate their genesis. The prevalent scientific consensus posits that galaxies emerge when dark matter coalesces and accumulates. This proposition is highly credible and aligns with my views.

The James Webb Space Telescope detected a minimum of five very large galaxies that appear to have formed much too quickly following the Big Bang.

As researchers, we employ several methods to determine how and when a galaxy formed, including Observational Data, Redshift Data, Stellar Population Data, and Cosmic Microwave Background Radiation.

However, there are instances where five particular galaxies appear to have formed unusually quickly, leading some Dimensional Anthropology Museum scientists to question if they are indeed galaxies at all. To address this, we need to approach the data with open-mindedness and fearlessness, allowing the evidence to guide our conclusions.

Hese unique galaxies exhibit intriguing characteristics, such as intelligence, the ability to combine and share, and defying the laws of gravity by moving away from each other.

By embracing the data without hesitation, we can gain a clearer understanding of the enigmatic phenomena at play within these peculiar celestial objects.

However, my research diverges from the conventional narrative in regards to the precise nature of dark matter. I propose that the images in this compendium have unveiled the intrinsic essence of dark matter. Dark matter, I posit, embodies the energy of life in its entirety. This is embodied in the phrase, "We are all star stuff."

Carl Sagan, the renowned astronomer and science communicator, popularized this phrase. It alludes to the empirical fact that numerous elements constituting our bodies, as well as all terrestrial matter, were engendered within the cores of stars via nuclear fusion. Over eons, these elements were disseminated throughout the cosmos via supernova explosions and other astrophysical phenomena, paving the way for the emergence of new stars, planets, and ultimately, life as we perceive it.

Sagan invoked this phrase to underscore the profound interconnectedness between humanity and the universe. However, I assert that Sagan's viewpoint did not extend far enough.

If these images are indeed authentic and provide evidence for continued existence, it elucidates our cosmic trajectory.

The James Webb Space Telescope, using infrared light, is capable of perceiving heat radiation, the residual light from the universe's inception. Infrared light, invisible to the human eye, typically renders images as areas of empty space. However, the reality is that space is teeming with dark matter, shifting and flowing between galaxies.

Upon discovering the profound images within a singular photograph from the James Webb Space Telescope, the Dimensional Anthropology Museum chose not to utilize the Iroquois Technology on additional photographs except for one.

To ensure the validity of the results obtained from the Iroquois technology, particularly concerning an image obtained via the James Webb Space Telescope, we employed a control method. We selected an additional random photograph to run through our analytical program. While the Dimensional Anthropology Museum has not received authorization to disseminate the full-color image to the public, we can present a cropped segment that has been revealed solely through our analytical processes. Preliminary analysis confirmed the program's accuracy when applied to this specific section of the image.

At first glance, the depicted structures could easily be misconstrued as wisps of diffuse gases. Yet, upon a more rigorous examination, we discerned an anomaly that we hadn't anticipated: the presence of what appeared to be an archaic helmet. Further assessment led us to posit that this artifact bore a strong resemblance to a helmet from the Trojan war era.

The subsequent observation of what appeared to be lipid blue ocular features within the helmet's eyeholes was both startling and unprecedented in our research.

One must wonder: Is this the spear of an Achaean? (Commonly referred to as Greeks during the Trojan War.) They certainly used spears, among other weapons, in their conflicts with the Trojans. In fact, the spear was a primary weapon of the ancient Greek warrior, or hoplite.

Homer's "Iliad" serves as our primary literary source for understanding the Trojan War. Within its verses, numerous scenes depict Greek and Trojan heroes wielding and throwing spears. This weapon was an essential component of the hoplite's armament, often accompanied by a large, round shield known as the aspis or hoplon. In battle, they employed the spear for thrusting in close combat and for hurling at distant foes. The implications of such a finding were profound enough to warrant an immediate re-evaluation of our study's objectives.

Provided below is the magnified section of the mentioned image, juxtaposed with a reconstructed representation of a Trojan helmet. A detailed inspection suggests significant battle-related wear on the helmet, especially a pronounced deformation on its right side.

This damage suggests a traumatic impact, perhaps from a heavy combat implement like a mace or battle mallet. The facial region beneath the nose guard seems to have suffered. While these entities might exhibit camouflaging abilities, this damage appears extensive. But what truly shook us to our core was the grim realization that the weapon believed to have ended this Trojan's life was gruesomely embedded, protruding from the back of his neck.

Figure 11a: Hector

28

In the course of our investigation, we employed advanced Dynamic Filtration techniques, revealing the presence of an arrow shaft centrally embedded within the anterior cervical region of the depicted Trojan individual. Subsequent analysis and comparison to extant artifacts from the period suggests that this arrow closely aligns with designs historically attributed to Greek archery during the Trojan War era.

Specifically, the arrowhead's metallurgical composition is consistent with looks like it could have been made from bronze; a material widely utilized for weapon manufacturing during this timeframe due to its accessibility and malleability. Detailed observation may also discern residual indications of fletching towards the arrow's distal end.

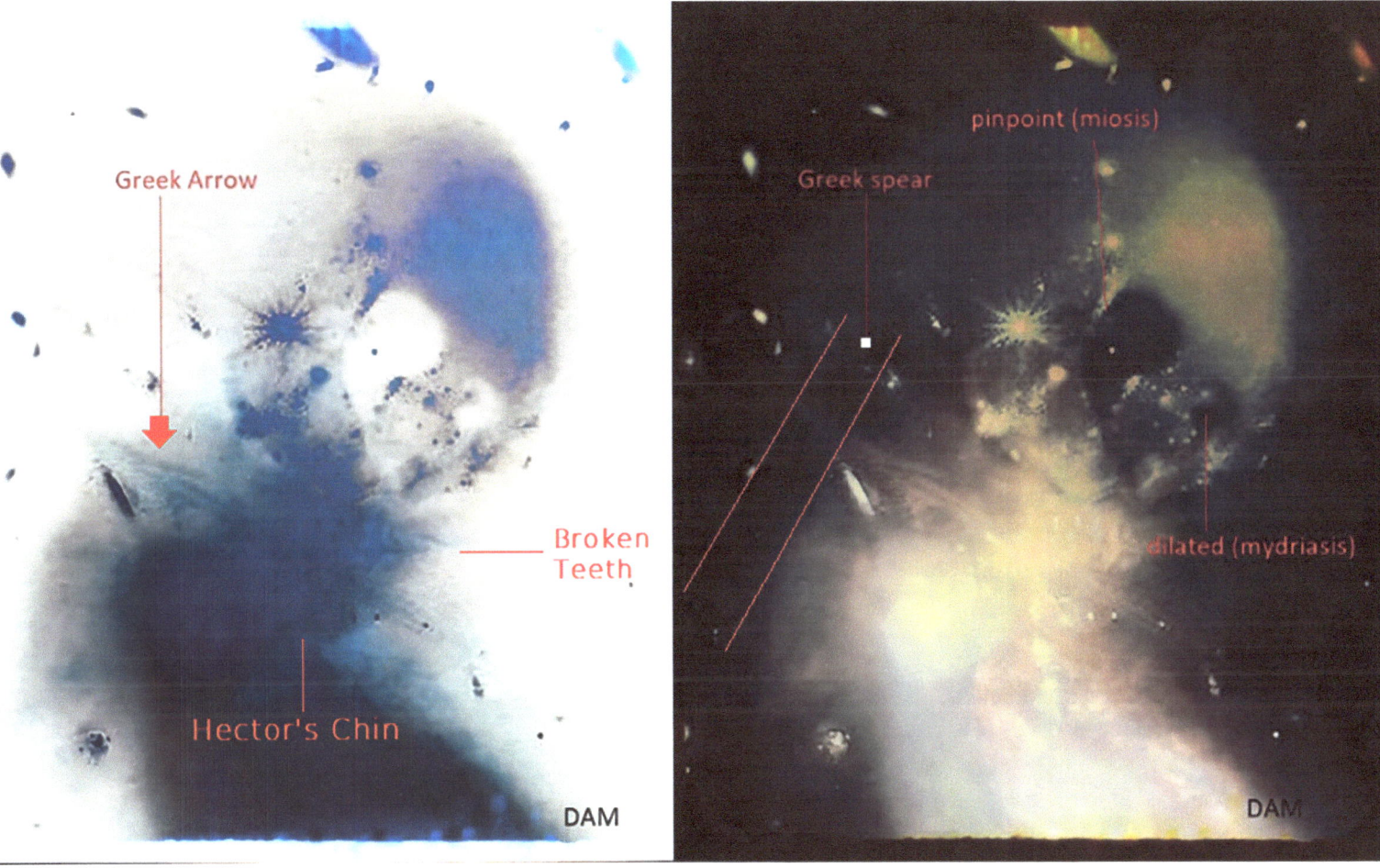

Figure 12: Hector

The more we looked at the images and figures Iroquois Technology discovered, we more and more started to wonder if the helmet Trojan was Hector, prince of Troy? The battle between Hector and Achilles, as described in Homer's ancient Greek epic poem, the "Iliad," is one of the most poignant and significant confrontations in Western literature. It represents not only a pivotal moment in the narrative of the Trojan War but also captures profound themes about honor, fate, and the human condition.

The "Iliad" tells part of the story of the Trojan War—a conflict between the city of Troy and the Achaean (Greek) warriors.

While it was sparked by the abduction (or elopement, according to some versions) of Helen, the wife of Menelaus, by Paris of Troy, the epic delves deeply into the characters, their motivations, and the nature of heroism and mortality.

Events Leading to the Duel:

Before the confrontation with Hector, Achilles had abstained from the war due to a personal slight by Agamemnon, the Achaean commander. He felt dishonored when Agamemnon took Briseis, a woman awarded to Achilles as a war prize. During Achilles' absence from the battlefield, the Trojans gained an advantage.

CHAPTER FOUR

However, the tide turned dramatically when Hector, with the help of Apollo, killed Patroclus, Achilles' close friend and comrade. Patroclus had donned Achilles' armor, hoping to inspire the Achaean troops and push back the Trojans. Hector mistakenly believed he had killed Achilles, took Patroclus' life, and claimed the armor for himself.

The Duel:

Driven by grief and rage over Patroclus' death, Achilles returned to the war, seeking vengeance against Hector. He had new armor crafted by the god Hephaestus and took to the battlefield with unmatched fury.

Hector, despite being warned by his parents and wife Andromache, decided to face Achilles outside the walls of Troy. As the two met, Hector proposed a pact: that the body of the loser would be returned unharmed to his people. Achilles, however, was consumed by rage and rejected this offer.

During their duel, Hector was initially protected by the armor of Achilles he had taken from Patroclus, but he was eventually struck down by Achilles, with a spear to the neck. It was at this very point when our researchers blood ran cold. As we looked on to the James Webb Space Telescope discovery aided by Iroquois Technology, it became Earthshakenly plausible that the Trojan warrior before us with the spear jammed deep into the back side of his neck could be Hector. We immediately sought out answers to the two eyes that could be seen within the eye holes of the helmet. The right eye where the spear had penetrated the neck. The pupils in the eye was pinpointed. Where the eye on the left and opposite side of the spear was dilated.

Our research showed that the neck houses numerous critical structures, including major blood vessels, nerves, the trachea, and the esophagus.

And if an injury occurred as significant as one involving a spear to the neck could cause a condition called Horner's syndrome, which is characterized by miosis. There is also two terms called "mydriasis" that refers to the dilation of the pupil. The opposite condition, in which the pupil is constricted or narrowed, is called "miosis. both conditions can result from trauma to the head and neck. If one studies the eyes seen within the helmet, the left eye appears mydriasis and the right eye looks miosis. It could be argued that Hector possessed similar conditions following and at the end of his life both these conditions. Below our team has generated a schematic representation delineating the positioning of wartime artifacts within the remains of the Trojan individual in question. The photograph to the left is the original image that Iroquois Technology hit on. The photograph to the right show the size, shape and location of the arrow, adjoined with the Greek spear to the neck that is not so clearly recognizable.

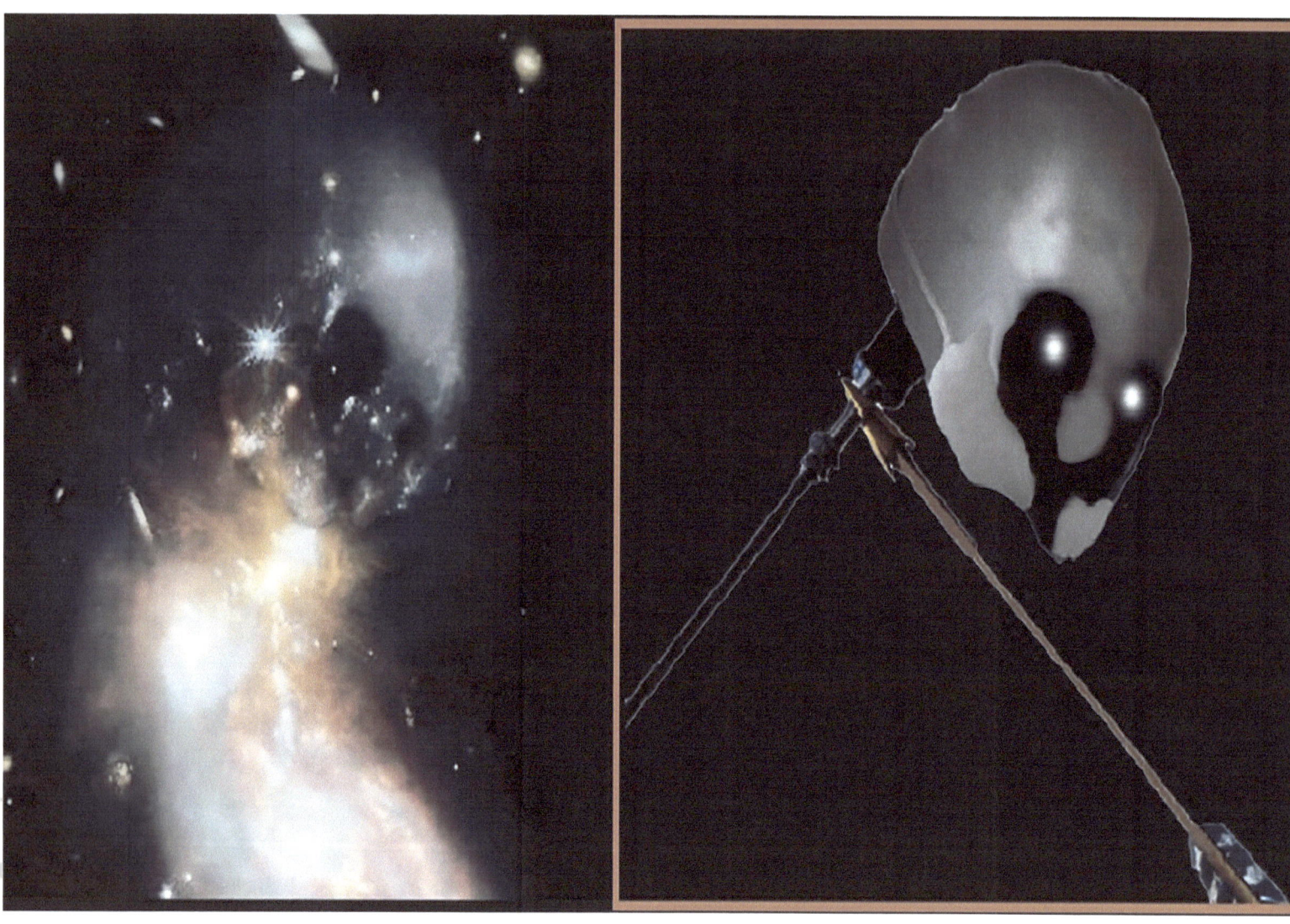

Figure 13: Hector

Whether real or imagined, this discovery far outranks the others, in my humble opinion.

It makes me realize that despite the technological marvels and advancements that characterize our modern existence, there is an intrinsic essence of human virtue established thousands of years ago that seems to have dissipated in a world dominated by contemporary devices, innovations, and groundbreaking medical achievements. We often find ourselves entangled in trivial disputes that warriors of antiquity like Hector and Achilles would deem unworthy of mention.

The appointed encounter between King Priam's prime minister of Troy and Achilles reminds us of a fundamental truth: even in enmity, respect can and should be extended. It was Priam's prime minister who told Achilles that enemies can show respect. Moreover, in those times, offering pardons to one's enemies was seen not as a sign of weakness but as a testament to the true strength and depth of one's character. This profound ethos appears to be a diminishing sentiment in today's society. Many in the present era mistakenly equate stubbornness with strength and perceive disrespect as a sign of virility. Yet, the men of yore displayed their valor, not through mere bravado, but by acknowledging the worth of their adversaries, extending due respect, and showcasing magnanimity through forgiveness.

It is this profound respect, rooted in recognizing shared humanity, that we seem to have misplaced in the corridors of time.

On the other end of the spectrum, this finding has hardened my heart to end the Iroquois Technology forever. This decision stemmed from the belief that further imaging from our programs would infringe upon what I like to call the privacy of God. We are a small, independent research organization, and the goal of our research was purely for entertainment. We did not create the Iroquois Technology to change or alter science in any way. We only wanted to use our science education to elicit better stories and mysteries for the world to enjoy.

Our goal was to use our programs to show the world that there was more to this universe and to share those findings. We never once attempted to convince the public that our findings were real; we only wanted to show what the program could unveil, and we left it up to individuals to determine what they considered real.

So, the members of the Dimensional Anthropology Museum understood that none of us were under any obligation to use the program in a way that might seem intrusive to what the program was designed to reveal. As the founder and CEO, following this discovery, I have decided to discontinue using the Iroquois technology indefinitely. This discovery has and always will serve its purpose for humanity. Now it's up to society to learn how to make the nation great again. It starts with respect for those we disagree with, and the initiative must come from us; we cannot wait for others to take the lead.

Those who shout the loudest for change must be the very ones who are willing to offer themselves up for the most significant changes to come. That is the only way to make a nation great. We do not sacrifice others in our journey for change; we must sacrifice ourselves for change to take hold. Change, real change that one demands must always start deep within us if we ask for modification. Without personal sacrifice, we are demanding others to do what we are unwilling to do to get the desired change. We must always ask what, and I am ready to sacrifice today for the change I seek for tomorrow? Without respect for our adversaries, mankind can't progress. Otherwise, we risk regressing to a time when grunting at one another was our only form of communication.

To clarify, I want to use this last statement to emphasize the importance of humility and reverence toward the vastness of existence. I urge others to appreciate the enchantment of what we do not fully comprehend.

This perspective resonates deeply with the idea that the pursuit of knowledge should include a sense of wonder and appreciation for the complexities of the universe. And with that, I have decided to abandon further exploration in this area.

The images we have seen in these photographs lead me to believe that something greater than us exists out there, and it is with trepidation that I have chosen to respect its right to peace. I understand that other research organizations may have their own decisions to make, and they may be bound by their stakeholders' or financial backers. I am not suggesting that they should stop or reconsider their impact; I can only speak for me and my research organization.

"On a side note, if I may veer off script for a moment: This discovery scares the bejeezus out of me!"

There, I said it, now lets continue.

We recognize as an organization that a significant portion, perhaps up to 90%, of the population may not discern the figures described in this publication. Moreover, among those who do perceive them, many might choose not to validate or acknowledge them.

For some individuals, personal priorities or beliefs may overshadow these observations. Acknowledging these images might, for some, challenge their prevailing perspectives. It's important to note that there is no concrete evidence to validate our claims. This explanation serves to share the reasons behind my research team's decision to discontinue the project.

Upon the publication of this paper, all records, notes, equipment, and evidence of the technology's capabilities will be deleted. I have implemented a failsafe into the programming to ensure that even I will not know how to recreate it.

For those who wish to dismiss the program as unreal and the images as easily explained away, I respect your views. We all have the freedom to interpret and question. Yet, for my research organization and myself, the implications are too profound to overlook. I am guided not by external skepticism but by a deeply rooted belief in a higher force beyond the realm of human comprehension.

Let us not forget that, by consensus, scientists worldwide recognize that we comprehend only about 5% of the universe's workings. This understanding is insufficient to definitively assert the accuracy or potential flaws in our research. By only understanding 5% of any topic or phenomenon means that there is a vast majority (95% in this case) that remains unknown or not understood. As a reminder to all, here are several reasons why such limited understanding is not conducive to making final judgments.

Limited Perspective: With only 5% understanding, our perspective is limited. Without a comprehensive view, our judgments could be skewed or biased.

Potential for Misinterpretation: If we only know a small fraction of something, there's a higher chance that what we do know might be misinterpreted because it's taken out of the larger, mostly unknown context.

Incomplete Data: Judgments or conclusions are best made when they're based on comprehensive data. Making decisions based on such limited understanding can lead to erroneous conclusions.

Dynamic Nature of Knowledge: What we understand about a topic can change as more information becomes available. If we've only scratched the surface of understanding, it's premature to make definitive conclusions.

Overconfidence: Believing that one can make accurate conclusions with only partial understanding can lead to overconfidence, which in turn can result in mistakes or misjudgments.

Missed Opportunities: If we make final judgments with limited understanding, we might close off avenues of further exploration that could lead to significant discoveries or insights.

In summary, making final judgments based on limited understanding is like trying to describe the entirety of an iceberg by only looking at its tip. The vast majority remains hidden beneath the surface, and without that complete knowledge, any conclusions made are inherently limited and potentially flawed.

So, let's not totally discard these finding flat out. As we await to discover the other 95% of the story that is still missing.

If Iroquois Technology did indeed discover the energy of the world's early inhabitants, this is where I choose to end the furtherance of this particular type of research by Iroquois technology. The mysteries of existence may remain veiled, and for now, I am content to let them be.

CHAPTER FIVE

"In our pursuit of understanding, let us not forget that some veils are meant to remain unlifted, for therein lies the beauty of awe and wonder." Quote by Researcher TD McRoy

If galaxies indeed embody what I hypothesize, they should remain undisturbed. Our posthumous existence will retain its enigma and will remain unaided by my revelation of these images, which will serve as a basis for generations of scientific inquiry.

As an astrophysics and AP Physics student, my knowledge remains embryonic. However, upon reading about the perplexing movement of galaxies away from each other at increasing velocities, defying conventional understanding of gravitational forces, an epiphany emerged.

What scientists are witnessing may well be a manifestation of the Heisenberg uncertainty principle. The tenet that the observation of particles can change particles, a foundational principle of quantum mechanics, could be the underlying explanation.

In the realm of quantum mechanics, particles can exist in a superposition of states until their observation 'collapses' them into a single state. The Heisenberg uncertainty principle posits an inherent limit to the precision with which pairs of physical properties can be simultaneously known.

This principle does not merely suggest that measurement devices influence particles; it indicates that the properties of particles are not definitively established until they are measured, and such measurements inherently involve a degree of uncertainty.

It's also important to note that this doesn't imply the conscious observation by a human or a living being is required, merely the interaction of the particle with the external world that could potentially provide information about its state is enough to cause the collapse of the wave function.

So in turn, the reason that the galaxies are moving in ways that seem unconventional to known laws of physics is because we are observing them. The Population Reference Bureau estimated that about 108 billion people had ever been born.

And if you start subtracting the current living population, that would be a lot of energy that needed to go somewhere. The answer to dark matter has been staring us directly in the face the entire time and no one figured it out. It's the First Law of Thermodynamics.

First Law of Thermodynamics: Often referred to as the law of energy conservation, it states that *energy cannot be created or destroyed, only transferred or changed from one form to another.* In other words, the total energy in an isolated system is constant. Meaning that what the James Webb Space Telescope is capturing are the energies of the earliest inhabitants of planet Earth.

Dear Esteemed Followers of Iroquois Technology and the Global Scientific Community, It is with a profound sense of gratitude and bittersweetness that I pen this farewell to you, the very individuals who have made my scientific journey a remarkable and illuminating endeavor. For nearly two decades, we have delved into the mysteries of our universe together, embarking on a shared quest for knowledge and truth. Today, I present my final contribution to this field, a publication that encapsulates the culmination of this extraordinary journey.

This final publication, resplendent with awe-inspiring images garnered from the annals of Iroquois Technology, represents a fitting denouement to my scientific odyssey. Henceforth, my energies and resources will be directed towards disaster relief, a realm where immediate and tangible impact can be made.

The objective of this publication is not to assert the existence of an afterlife or to transform the current scientific paradigm. It exists as a testament to our shared curiosity, a token for those captivated by the intriguing topic and mesmerizing images it presents. Fundamentally, this book is designed to serve as a source of enlightenment and enjoyment, a bridge between the realms of science and entertainment.

The Iroquois technology has rendered a rich tapestry of incredible images over the past seventeen years. However, the recent discovery within the stars marks the terminal point of this technological venture.

It is crucial to understand that at this juncture, these images remain within the realm of entertainment rather than rigorous scientific evidence. The world is not yet primed to embrace these extraordinary images as concrete scientific fact. The current global landscape calls for a reaffirmation of our shared humanity and compassion, a collective return to the principle of loving thy neighbor.

These images, now immortalized within the pages of this book, will serve as a repository for future generations of researchers. For now, they will continue to spark intrigue, stir imaginations, and provide a source of entertainment.

As I bid farewell to this chapter of my life, I express my deepest gratitude to each of you for accompanying me on this journey. Your support and engagement have made this expedition into the unknown a truly enriching experience.

Onward to the next chapter and the next adventure! In my final years on this planet, I am dedicated to helping those in need. To achieve this, I have founded a nonprofit 501(c)(3) organization called The Bertha Project – Dimensional Anthropology Museum. The goal is to establish a more efficient and effective outreach program that can swiftly deliver food, water, and other essentials to those affected by man-made and natural disasters within hours, not days.

Our primary vehicles, named Bertha & Kiki, are equipped to serve this purpose. They will be staffed by retired firefighters and law enforcement officers, operating similar to a volunteer fire department, always prepared to deliver essential goods directly to communities in need. Just like an ice cream truck, we will go street by street, providing life-saving supplies free of charge. This approach greatly benefits the elderly, people with mobility issues, or those caring for sick loved ones, as they can receive supplies at their curbside without having to wait in long lines or carry heavy items long distances.

The Bertha Project anticipates the future challenges as communities continue to face devastation. If you wish to support our efforts, please consider donating on our website, www.theberthaproject.org. The proceeds from the sale of this publication will contribute to the construction of the Bertha & Kiki trucks', capable of servicing around 250 homes before the need to restock Bertha & Kiki. Your donation is greatly appreciated, as it can make a life-saving difference.

Also, to assist our 501(c)(3) in raising funds for disaster relief efforts, we are considering a drawing for a signed copy of this book. Only one copy will ever be signed by the author, TD McRoy. No other copies will bear his signature. As the author, he has chosen not to sign additional copies because he genuinely believes that God inspired his research and feels claiming credit for more than one copy is more than he deserves. Offering one signed copy to a fan, follower, or supporter of this project is more meaningful than affixing my name to hundreds of books that might not hold significance to many in the end. Such a unique book becomes treasured by its owner for a lifetime.

As I conclude our discourse, permit me to unveil another celestial masterpiece, a fresco painted by the heavens themselves. This exquisite work captures the essence of the universe's intrinsic artistry—a Native warrior, resplendently adorned, emblematic of the profound potency embedded within the stars. Such celestial creations remind us of the imperative to cultivate an unwavering reverence for the cosmic wonders that guide our earthly voyage.

In closing, I am pleased to share four research papers I penned subsequent to the transformative revelations that reshaped my fundamental perceptions of the cosmos. These manuscripts delve into the depths of the discoveries that have forever altered my understanding of the universe we inhabit. The photograph adorning the adjacent realm of this discourse presents but a minuscule fragment of the panoramic vista unveiled by the inquisitive gaze of the James Webb Space Telescope.

The cropped image below, on the right is the exact photograph that highlights the visage of the great warrior. In the heart of this untamed space, where ancient existence seems to intertwine with the rhythm of nature's expanding universe, there exists a magnificent marvel—a large native headdress, resplendent with an enchanting array of beautiful feathers. Like the majestic wings of a soaring eagle, this exquisite headpiece bears testament to the harmony between man and nature. The image speaks of unparalleled mastery, as each feather is artfully arranged, creating a mesmerizing tapestry of colors that reflect the kaleidoscope ca3st1 among the stars.

The rich tones mingle with the vibrant hues of wisps of energy flows, weaving together a captivating story of unity and diversity held amongst the heavens.

One can only surmise that this headdress symbolizes the inherent connection between the spiritual and the earthly realms. It is a bridge that spans across cultures and generations, speaking the language of heritage and wisdom.

As the headdress graces the noble wearer, it brings forth a whispering of ancient tales of bravery and resilience, echoing the courageous journeys of those who came before him.

This image, in particular, seems to stir my soul, and it ignites a profound sense of belonging to the scale and the grandeur of what makes us who we truly are or who we should be. It suggests that Earthly squabbles of politics, power, money, and greed are so insignificant when looking upon the true power of the universe—it is truly awe-inspiring to behold.

Figure 14: Native Warrior

The discovery of celestial anomalies through Iroquois technology presents an enticing opportunity, as new galaxies emerge from the depths of the cosmos. The potential for lucrative gains in unraveling the mysteries of the universe is undeniable. However, amidst the allure of wealth and prestige, one must remember the core principles that guide us—the reverence for nature's grace, the humility before the divine, and the preservation of the enigmas that shroud the heavens.

In the pursuit of knowledge, we must recognize that there are realms beyond human comprehension, domains where the hand of God and the intricacies of nature intertwine. The honor we owe to God, humanity, and the natural world must always take precedence over the pursuit of riches. It is within these values that we find our true compass, ensuring that our decisions remain noble and virtuous.

While the allure of exploration and discovery beckons, it is crucial to tread carefully and respect the boundaries that define our connection with the cosmos.

As I reflect upon my own research and the strides made through Iroquois technology, I recognize that there lies a boundary that should not be crossed—a boundary that safeguards the sanctity of the heavens and the mysteries that reside within.

As a custodian of this knowledge, I stand firm in my conviction that should science ever transgress that boundary, I shall not offer my assistance or the aid of Iroquois technology. There comes a point where exploration should not be driven by ambition alone, but rather tempered by the wisdom of knowing when to halt our steps and let the mysteries be.

Upon reaching the final zenith of discovery, an overwhelming sense of bliss and contentment engulfed me. The satisfaction derived from honoring the sanctity of the heavens and remaining true to the research's pure intent surpasses any amount of wealth that might beckon.

The hidden treasures within undiscovered galaxies shall remain untouched by my hand, for the pride lies not in the pursuit of riches, but in the fidelity to the sacred path of knowledge.

Regarding the images and the program that facilitated their discovery; with global accessibility now achieved, my words may fall short in encapsulating the depth and significance of what is portrayed. Henceforth, I have decided to maintain public silence on the matter, considering it has already been articulated through written means.

CHAPTER SIX

I shall entrust individuals with interpreting the essence of these images, refraining from further elaboration or engagement, allowing their personal perceptions to flourish without influence from me or my organization.

Let me share a fewearlier research papers I reconfigured following the analysis of the James Webb Space Telescope photographs. The discoveries have altered and fine-tuned many of my earliest learnings in the study of space. I have published an updated version for you to enjoy.

Research paper number one is paramount for understanding the universe's structure, evolution, and dynamics. Historically, various methods, including parallax and standard candles, have served as benchmarks. However, as with any scientific model, these methods are derived from observations and assumptions that can be subject to limitations or oversights.

Research paper One:

The Dog Tail Analogy:

Imagine an experiment wherein sensors are affixed to a dog's paws to measure the distance it travels. Based on its steps in a controlled environment, researchers deduce the dog has traveled 10 miles. This measure becomes an established standard.

In a subsequent experiment, researchers aim to calculate the distance a dog would need to travel to catch its tail. Using the aforementioned method and without direct observation, they deduce the dog has traveled 10 miles overnight. The logical conclusion would be that the tail was 10 miles away.

However, in reality, the dog's tail was merely inches away; the dog had merely run in circles before eventually capturing its tail after exhaustion.

Figure 15: The Dog Tail Analogy

Implications for Astrophysics:

By analogy, the Rover Theory postulates that our methods for calculating galactic distances might be analogous to the dog running in circles. While galaxies might appear vastly distant based on our established calculations, they could, in fact, be much closer than perceived. The current models might capture not only the linear progression of a celestial body but also its intricate movements, which could be analogous to the dog's circular chase.

Discussion:

One implication of the Rover Theory is the necessity to reevaluate our understanding of cosmic structures, dynamics, and interactions. If galaxies are closer than previously believed, this could reshape theories about dark matter, galactic interactions, and the overall fate of the universe.

It is essential to approach this theory with caution, recognizing the monumental achievements and rigor of past methodologies. However, the Rover Theory serves as a reminder that scientific understanding is continually evolving. It underscores the importance of questioning established paradigms and being open to novel interpretations.

Conclusion:

The Rover Theory invites astrophysicists to rethink established methods of measuring galactic distances. Just as the dog's tail was closer than the calculated distance suggested, galaxies might be nearer than current models indicate. As science progresses, it is paramount to ensure that our methods and interpretations align closely with the intricate realities of the universe.

Research paper two:

Are Galaxies Closer Than Perceived?: The Hourglass Projection Theory

Abstract: This research paper introduces the Hourglass Projection Theory, postulating that galaxies' perceived vastness might be an illusion, akin to our inability to perceive subatomic particles without magnification. The theory explores the role of black holes as potential galactic projectors and suggests that the vast cosmic distances we perceive might be a carefully crafted optical illusion. *

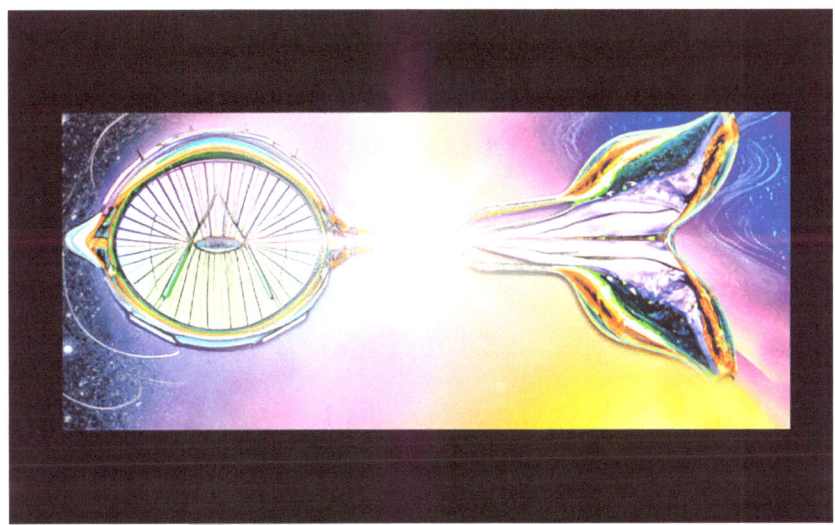

Figure 16: The Hourglass Projection Theory

Introduction

With state-of-the-art telescopic imagery, particularly from the James Webb Space Telescope, distant galaxies are showcased as vast cosmic entities. This leads to an essential query: if these galaxies are closer than they seem, why aren't they easily discernible without advanced technological aid?

The Subatomic Parallel: A perspective from the subatomic world can elucidate this conundrum. For instance, a foot affected by a bacterial infection. Despite the proximity of the observer's eyes (a mere 4-6 feet), the bacteria remain imperceptible. This oversight is not about distance but about scale and perception.

The Hourglass Analogy: Consider the dynamics of an hourglass. When observing through its constricted center at an image on its broader end, the image appears enlarged. Similarly, when our telescopes probe the cosmos, they might be capturing an enlarged reflection of much smaller cosmic structures.

Historical scientific standards might inadvertently anchor us to this magnified perspective.

Black Holes as Galactic Projectors: Central to this theory is the transformative role of black holes. Rather than simply cosmic vacuums, they may function as galactic projectors.

47

They might be casting the universe onto a vast backdrop, creating the illusion of expansive space. We, as cosmic inhabitants, are then beguiled into believing in these vast distances.

The "Moving Goalpost" Phenomenon: As our terrestrial technology and understanding burgeon, our aspirations to access these ostensibly remote galaxies escalate. However, the 'projection' might adjust, ensuring that the cosmic canvas remains elusive. This "moving the goalpost" phenomenon indicates that as we extend our cosmic reach, the projection undergoes continual recalibrations.

Conclusion: The Hourglass Projection Theory posits a potentially bounded spatial environment, challenging the notion of an infinite cosmos. Space, in its remarkable intelligence, constantly places the cosmic theater slightly beyond our comprehension. Leveraging the coherent attributes of laser light, with synchronized photons creating a potent beam, it's plausible that minute galaxies are magnified to appear vast and distant. The proximate location of the Canis Major Dwarf Galaxy, relative to the Milky Way, further strengthens this hypothesis. The galaxies detected by the James Webb Space Telescope could, theoretically, be no more extensive than an average human. These galaxies, possibly projected by black holes, might be cast onto a canvas reminiscent of an infinity mirror, yielding endless reflections. Causing a galactic mirage we all fell victim to.

Research paper three:

Black Holes—Outer Space Camouflage

For those seeking deeper insight into The Dimensional Anthropology Museum's theory about the purpose of black holes in our universe, we have crafted a celestial chart that aids in visualizing the intricate process and underlying purpose. Our journey commences on the left side of the chart, gradually moving rightward, unveiling a narrative that seamlessly blends cosmic phenomena with profound revelations.

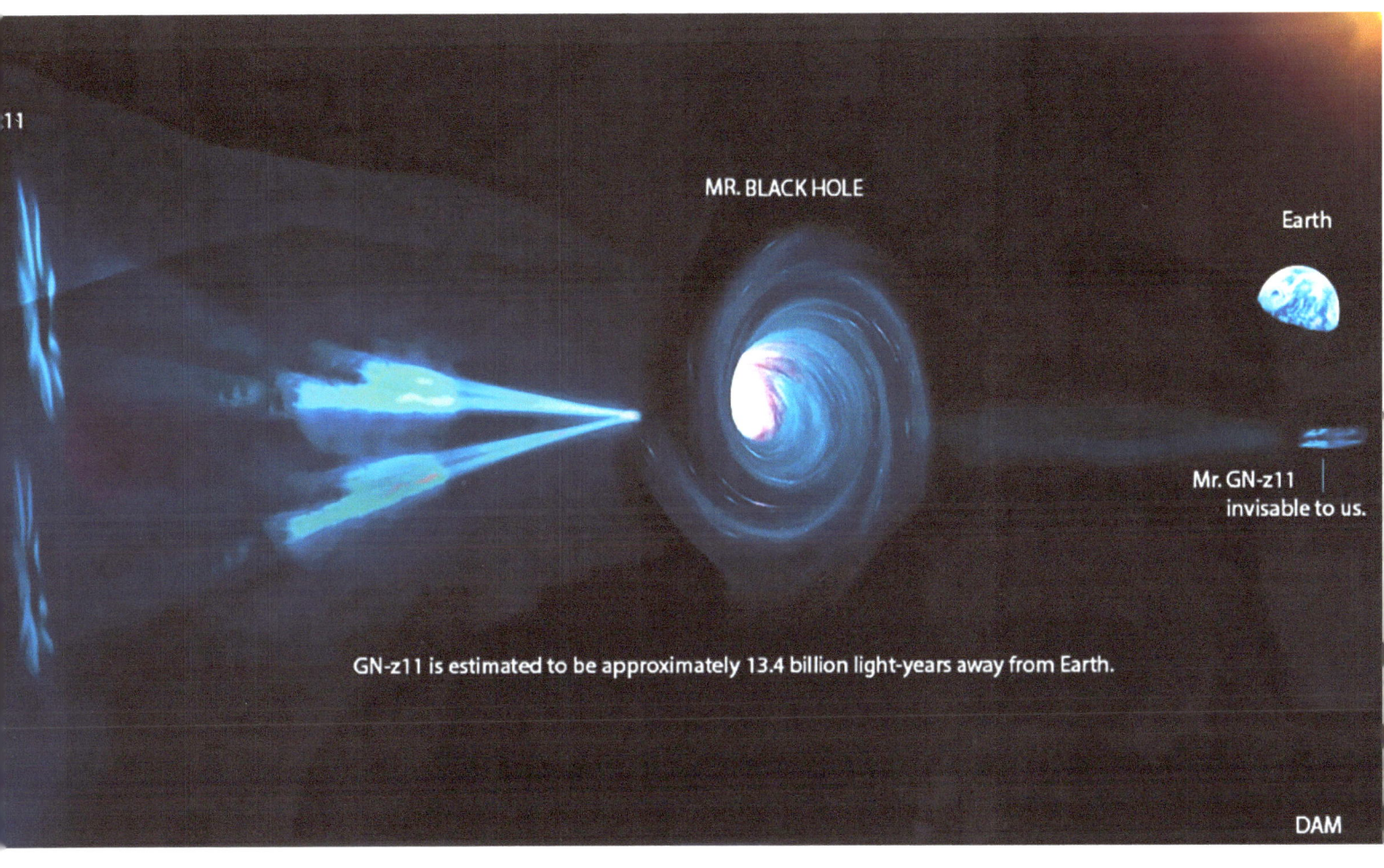

Figure 17: Black Holes—Outer Space Camouflage

Initiating our voyage, we're greeted by an intricate array of light patterns that gracefully resemble the petals of a celestial daisy. These captivating patterns artistically represent the GN-z11 galaxy, a contender for the title of the farthest known galaxy from Earth. This enigmatic galaxy, nestled within the Ursa Major constellation, was unveiled to the world through the meticulous observations of the Hubble Space Telescope. Its discovery has captivated astronomers and enthusiasts alike.

Continuing our traverse to the right, a gravitational giant takes center stage – a colossal black hole. This central entity, characterized by its immense gravitational pull, functions akin to a cosmic vacuum, ensnaring any matter that ventures too close to its elusive event horizon.

Our expedition guides us further rightward to the familiar presence of Planet Earth, a fragile blue jewel adrift in the vast cosmic sea. Beneath Earth's dominion emerges a celestial counterpart – a diminutive galaxy that intriguingly shares its name with the earlier-mentioned GN-z11 galaxy. However, this apparent repetition carries a hidden significance that warrants deeper exploration.

In a profound twist, the lower Mr. GN-z11 isn't an autonomous galaxy but the same galaxy as GN-z11. The GN-z11 seen on the chart far left is only the reflection of Mr. GN-z11. The 13.4 Million light-year galaxy is only a mirror reflection of Mr. GH-z11 who is still located just below Earth. It's all a mirage. This elusive entity, akin in magnitude to an invisible 6-foot man, exists in proximity to Earth's gravitational sphere. Mr. GN-z11 and his ethereal associates collectively constitute approximately 95% of the universe's matter. Their enigma lies in their intangibility, evading direct visual observation.

At the chart's heart emerges an imposing figure of the – Black Hole. This entity, characterized by its potent gravitational grip, draws in matter daring enough to approach its formidable event horizon. Intriguingly, the matter being pulled into this cosmic maw isn't the elusive Mr. GN-z11 and his intangible companions dispersed throughout space. Instead, The Black Hole engages in a unique interaction with the reflections of Mr. GN-z11 and his mysterious allies.

These enigmatic reflections, invisible to human senses, have long eluded our observation.

Yet, recent technological strides, epitomized by the James Webb Space Telescope, have enabled us to capture and decode the reflections of Mr. GN-z11 and his imperceptible retinue, originating from the cosmic depths.

CHAPTER SEVEN

To decode the riddle of the Black Hole's rapid rotation, our researchers turned to an unexpected source – a "Bell and Howell 1580 16 mm projector." This seemingly disparate device unearthed startling parallels between its mechanics and the hypothesis enveloping the Black Hole. This revelation spurred the expansion of our theory, triggering an illuminating realization.

In essence, the Black Hole is pilfering the invisible reflections of tiny galaxies and casting their images millions of light-years away, much like shining a flashlight on a tall building and observing the beam enlarge as it ascends the side. Our researchers began to ponder: if all this holds true, what impels space to behave in this manner? What drives this purpose? We contemplated that it might simply be the inherent evolution of the universe. The utilization of camouflage is observed in Earth's nature, so is it far-fetched to believe that space possesses its own method of concealing itself? We discovered that nature employs several distinct forms of camouflage.

Background matching: This prevalent type of camouflage in nature involves plants with coloration that matches their background. The living stone plant, for instance, has a grayish-green hue that enables it to blend seamlessly with the rocks it grows on.

Disruptive coloration: This variety of camouflage breaks up a plant's outline, rendering it harder to discern. The passion vine, marked by its mottled pattern, serves to disrupt its outline effectively.

Countershading: This camouflage technique entails making a plant's upper portion darker than its underside. This confounds herbivores viewing from above. The cactus, with its dark green top and light green bottom, remains challenging to spot when erect.

Mimicry: Certain plants mimic other objects, like rocks or sticks, to evade detection. The dead man's fingers plant, with its brown, stick-like appearance, effortlessly blends with the ground.

Camouflage proves pivotal for plants and animals, and now, according to our theory, it might be essential in outer space as well. If accurate, our theory offers a new dimension of understanding, aligning celestial behavior with the principles that govern life on Earth. Furthermore, we propose a compelling reason: the universe might be attempting to hide itself from view. Some readers might take issue with our theory, but it's a possibility we strongly believe should be considered.

Why Outer Space Might Use Camouflage: The Great Deterrence Theory

For millennia, human beings have gazed upon the vastness of the universe, pondering its limits and wonders. But what if the cosmos has been gazing back, and even more intriguingly, hiding its most profound secrets from us? Researchers from the Dimensional Anthropology Museum propose a radical theory: space camouflages its true extent to protect other celestial bodies from Earth's most invasive species – humans.

The Universe as a Sentinel

First, let's establish a foundational premise: the universe is a dynamic and sentient entity, with the capability to recognize threats and act in self-preservation. According to researchers at the Dimensional Anthropology Museum, over billions of years, the cosmos has had experiences with humankind, observing our actions, tendencies, and more alarmingly, our insatiable appetite for resources.

It's clear why the cosmos might deem us a threat. Our history tells a tale. Human beings are often likened to locusts, consuming resources without foresight and leaving a trail of environmental destruction.

From deforestation to the extinction of countless species and the dramatic effects of climate change, we've consistently showcased our capacity to exploit and damage our home planet.

Now, project this behavior onto a cosmic scale. As our technological advancements progress, the aspiration to colonize other planets becomes more potent, potentially leading us to mine these planets for resources, alter their ecosystems, and possibly render them uninhabitable, as we've done to numerous ecosystems on Earth.

Cosmic Camouflage

The researchers at the Dimensional Anthropology Museum suggest a thought-provoking hypothesis: What if space, in its profound wisdom, has preemptively devised a strategy to deter us? The notion of cosmic camouflage emerges here. Analogous to how animals on Earth have evolved to blend with their environments as a defense against predators, space might deploy camouflage to conceal its true expanse and the bounties of worlds far beyond our grasp.

Could the vast voids we believe exist between galaxies merely be deceptions, crafted to dissuade us from journeying too far? Perhaps there are planets bursting with life and untapped resources, but they remain concealed behind cosmic illusions, impervious to our current technological instruments.

A Protective Mechanism

This idea of cosmic camouflage transcends mere self-preservation. It could serve as a universal shield. If humans from Earth truly embody the "locusts" of the cosmos, the universe has every reason to prevent us from accessing its most susceptible worlds, thereby preserving them from prospective harm and exploitation.

Conclusion

While this theory, backed by the Dimensional Anthropology Museum, verges on science fiction, it casts an enthralling light on our interrelation with the cosmos. The crux might not be about validating whether space conceals itself from us, but rather understanding why it might opt to do so.

It's a cosmic summons for humanity to introspect, evolve, and aim to be more responsible inhabitants.

To pose a thought: "Let's be honest. If you inhabited a vibrant, untouched planet with boundless resources that have never been mined or fracked, would you desire humans to step foot on this paradise? This is the paramount question we ought to address collectively. Are any of us truly worthy to be trusted with a virgin planet?

Another possibility could be what we call Camouflage in the Cosmos.

Camouflage in the Cosmos: How Outer Space Might Be Concealing Its Secrets

The vast expanse of space, with its inky blackness punctuated by bright pinpricks of stars and galaxies, has always been a source of endless wonder and mystery for humanity. But what if, hidden within this vastness, are secrets concealed by the universe's own form of optical camouflage? ### The Fresnel Lens Principle At the heart of this discussion is the concept of the Fresnel lens. A Fresnel lens is a special type of lens that is flat on one side and ridged on the other.

It focuses light in a manner similar to a regular convex lens but does so over a much shorter distance. The idea behind the lens is its ability to bend and redirect light, creating optical illusions or focusing light in specific ways. Imagine, for a moment, if space itself behaved in a manner akin to a massive Fresnel lens.

Such a phenomenon would mean that light from distant objects, or even entire galaxies, could be bent or redirected in ways that make them appear differently or perhaps even remain entirely hidden from our observation.

Cosmic Camouflage: How Might It Work?

1. **Gravitational Lensing: ** We already have evidence of a natural phenomenon similar to the Fresnel lens principle called gravitational lensing. Massive objects like galaxies and black holes can warp the fabric of spacetime around them. This warping causes light from objects behind them to bend around, often leading to distorted or multiplied images.

2. **Cosmic Mirages:** Just as heat waves on a hot road can create the illusion of water, vast fields of varying temperatures and densities in space could potentially distort the light, making certain areas or objects appear differently or making them invisible.

3. **Quantum Fluctuations:** On the quantum scale, particles and antiparticles pop into and out of existence. This quantum "froth" might, under the right conditions, affect the passage of light in unpredictable ways, causing unexpected obscurations or distortions.

The Implications of Cosmic Camouflage If space truly possesses its own natural forms of optical trickery, the implications are profound:

1. **Hidden Celestial Bodies:** Stars, galaxies, or even vast clusters might elude our detection if their light is being diverted away from our line of sight.

2. **Mysteries of Dark Matter and Dark Energy:** These elusive substances that seemingly make up most of the universe's mass and energy remain a mystery. Could it be that some of their effects or even direct evidence of their existence are hidden behind cosmic optical illusions?

3. **A New Understanding of the Universe:** Recognizing and accounting for these camouflaging effects could radically alter our understanding of the universe's size, shape, and overall structure.

In Conclusion While it's a tantalizing thought, the idea that space uses a kind of Fresnel lens-like camouflage remains in the realm of speculative science. Yet, it's these sorts of imaginative leaps that often lead to groundbreaking discoveries. As we refine our observation tools and techniques, we may one day unveil the universe's grandest magic tricks and finally uncover the secrets it has so masterfully hidden from us.

Research paper four:

** Is Time Travel Theory Dead? **

Title: "Observational Time Observance (OTO): Rethinking the Foundations of Time Travel"

Abstract:

This research paper presents a profound departure from conventional time travel theories, offering a fresh perspective inspired by the insights of luminaries such as Albert Einstein and H.G. Wells. Leveraging the captivating imagery captured by the James Webb Space Telescope, this study introduces a groundbreaking concept named Observational Time Observance (OTO). This innovative framework challenges the feasibility of traditional time travel notions while proposing an alternative approach to experiencing past events. Anchored in the intricate interplay between historical occurrences and the omnipresent Higgs field, the OTO theory posits that energy and memory can be accessed through specific technological means. This paper implores the scientific community to embrace a paradigm shift, casting aside the limitations of conventional time travel theories and embarking on a new trajectory of understanding.

CHAPTER EIGHT

Figure 18: Observational Time Observance (OTO)

1. Introduction:

Throughout history, the tantalizing concept of time travel has captured human imagination, drawing inspiration from visionaries like Einstein and Wells. As we delve into the mesmerizing vistas revealed by the James Webb Space Telescope, our very notions of time travel undergo a metamorphosis. This paper introduces a revolutionary concept—Observational Time Observance (OTO)—contending that traditional time travel concepts, whether forward or backward, face inherent improbabilities. Instead, a novel model emerges—one that places Observational Time Observance at its heart. This model steers away from intricate machinery, favoring a technologically pragmatic approach rooted in temporal observation.

2. The Higgs Field and the Essence of Time:

At the core of this revolutionary hypothesis lies the intricate interplay between past events and the enigmatic Higgs field—an elemental fabric that weaves through the cosmos. Analogous to how smoke particles intermingle with fabric, historical events imprint their essence as energy within the Higgs field. This enduring energy, echoing the tenets of thermodynamics, endures beyond time's constraints. Observational Time Observance suggests that these lingering energy imprints can be activated and harnessed through a technological mechanism reminiscent of the Iroquois Technology employed by the Dimensional Anthropology Museum.

3. A Radical Approach to Experiencing the Past:

Observational Time Observance introduces a visionary device, adept at rekindling dormant energy imprints—providing access to historical events without traversing time.

By eliciting these residual energy imprints, specific past events can be re-lived within their original contexts. This paradigm departs from the traditional portrayal of time travel found in science fiction, placing emphasis on grounded, reasoned observation.

Figure 19: Observational Time Observance (OTO)

4. Challenges and Implications:

While the potential of Observational Time Observance is captivating, it necessitates surmounting significant challenges. Foremost among these challenges is the development of a mechanism to activate dormant energy imprints. Manipulating Higgs field energy demands inventive ingenuity and collaborative efforts among physicists and astrophysicists. Should this endeavor bear fruit, it promises to herald a transformative shift in how humanity perceives and engages with its historical narrative.

Therefore, our proposition suggests that instead of embarking on space voyages or employing terrestrial mechanisms, this apparatus would enable an observer to spectate a historical occurrence precisely as it unfolded, spanning back centuries or even millennia.

Analogous to the lingering aroma of a cigar in a mansion long after the original owner's departure, our theory posits that the residual particles of bygone times, much like images imprinted in the ethers of memory, persist. We only need to unlock those particles.

5. A New Epoch of Understanding:

In closing, this research champions a paradigm shift, inviting scholars to relinquish traditional time travel concepts in favor of embracing Observational Time Observance. The resplendent imagery unveiled by the James Webb Space Telescope beckons us to extend the frontiers of human exploration. This paper aspires to embolden scientists to adopt a pragmatic stance, charting a course toward comprehending the intricate interplay of energy, memory, and the ubiquitous Higgs field—a journey that promises to illuminate the intricate fabric of our universe through Observational Time Observance (OTO).

By the time we researched the end of this scientific journey, the information that was gathered foretells of a time of Greek Mythology and the possibility that mankind's knowledge on the topic had somehow been bred out of us. The learning from Homer's Odyssey was profound and led me to use the information discovered by this evidence to provide Homer with a final chapter and a conclusion. God rarely uses the rich and powerful to tell his stories; He uses the most humble and ordinary of individuals to deliver his messages to the future of humanity. For this reason, I have added a tale as a way to sanctify this research and connect entertainment with science, the initial essence upon which the Dimensional Anthropology Museum was created. Below is how we envision the story's end for mankind and the salvation of our planet.

The conclusion of Homer's Iliad: Elysium's Retribution

In the vastness of the cosmos, among constellations and interstellar dust, lies the Elysium Realm, the fabled land of the dead where gods, heroes, and mythical beings reside in spirit form. These celestial entities, once rulers of Earth's landscapes in Greek tales, were banished to the stars, transforming space into their sanctuary.

Odin, with his single piercing eye, surveyed the universe alongside Zeus, the ruler of the Olympian gods. Together with an assembly of deities and heroes, they formed an alliance that watched over Earth from the great cosmic divide. Their exile to the stars had not diminished their power, but it had rendered them undetectable to mortal eyes. From the Earth, space appeared as an endless expanse waiting for exploration. With dreams of expanding their horizons, nations competed in the great Space Race, launching rockets, probes, and satellites.

But unbeknownst to mankind, every fiery ascent was perceived by the celestial beings as an act of aggression.

The spirits of ancient Trojans and Greeks, once sworn enemies, now stood shoulder to shoulder in Elysium. Their united front wasn't formed by a shared history but by a collective defense against Earth's encroachments. Every rocket's blaze and every satellite's orbit was an unintended slight, a disruption to their ethereal peace. These factions of Earth were headed by clans called NASA, CNSA, ESA, Roscosmos, ISRO, and JAXA responsible for firing the first shots into their realm.

As the temperature of Earth began to rise, a result of mankind's insatiable consumption of fossil fuels, the celestial realm saw an opportunity. The very elements which the gods once wielded to display their displeasure were now being supercharged. Hurricanes, earthquakes, tsunamis — nature's rage was magnified, becoming powerful tools in the hands of the divine.

The legends of old had tales of gods sending disasters as punishments, but the people of Earth, consumed by science and reason, had forgotten. What if the modern world's climatic upheavals were not just natural consequences but orchestrated by cosmic entities? What if the sweltering heat, the rising oceans, and the devastating storms were celestial retaliations for the perceived intrusions?

Hidden within the data streams of the James Webb Space Telescope, technology detected unusual patterns. Ethereal figures, traces of godly auras, and ancient symbols were concealed within nebulae and star clusters. If the frescos of celestial beings weren't just myths; then it could be assumed they were very real and very displeased.

Yet, the heart of this cosmic conflict was a tragic misunderstanding. Earth's nations sought to touch the stars, to understand the universe, driven by curiosity rather than conquest. The rockets weren't weapons, but vessels of exploration, symbols of humanity's insatiable thirst for knowledge.

Zeus, in a council with Odin and other deities, voiced his concerns, "These mortals, in their blind pursuit, have waged war against us?

Odin, wisdom evident in his gaze, responded, "It's war they seek, and they have unknowingly provoked the heavens.

A great conundrum faced the celestial realm: To continue their wrath, forever condemning mankind for their unintentional transgressions, or to seek a bridge of understanding, to illuminate the darkness of ignorance with the light of knowledge.

In the great amphitheater of the stars, a decision was reached. Hermes, the messenger god, would descend to Earth, revealing the truth of the celestial realm that modern inhabitants of Earth have long forgotten.

It was time for those who have ravaged the planet to stand before justice of a new order and one of great power well beyond mortal man wheeled.

And so, as the Earth gazed upwards, filled with fear and wonder, they were met not with the silent aroura of distant stars but with a message from the very gods they had long forgotten. The next era of exploration wouldn't be marked by rockets and rovers but by reconciliation and respect.

In the tapestry of time, the story of the Celestial Wars stands as a testament to the importance of understanding and the peril of ignorance. It reminds all of humanity to tread lightly, to seek knowledge, and to remember that the universe, in all its vastness, is a realm of mysteries, tales, and ancient beings.

The Celestial Wars: Reconciliation of Realms

Upon Hermes' second descent to Earth, his ethereal form took on a tangible glow, illuminating the skies. His presence caught the attention of all, from young children looking out of their windows to leaders gathered in emergency summits. His resonant and soothing voice echoed through every corner of the world.

"Children of Gaia," Hermes began, "For eons, the cosmic divide has separated our realms. Your explorations, though born out of innocent curiosity, have reached the sacred sanctuaries of gods and heroes."

World leaders, scientists, and spiritual heads convened. At a summit in Athens, beneath the shadow of the Parthenon, Hermes delineated the grievances of the celestial realm. He spoke of the cosmic balance disrupted by Earth's intrusions, but he also acknowledged the pure intentions of humanity.

However, it wasn't just words that Hermes brought. He offered a solution—a bridge that would connect Earth and Elysium, not through rockets or probes, but through understanding and mutual respect. The Oracle of Delphi, once thought to be a relic of the past, would be rekindled as a conduit between realms.

CHAPTER NINE

To cement this newfound alliance, representatives from Earth were chosen to visit Elysium. Among them were scientists, artists, and philosophers—those whose pursuits were devoted to the betterment of mankind rather than self-interest. One who could grasp the depth of the cosmic dance and convey it to the masses.

Their voyage was imbued with a heart-stirring mystique. Guided by the ethereal Pegasus, they soared amidst constellations, their souls touched by the sheer splendor of the heavens. Standing in the hallowed presence of the Council of Gods, where Odin and Zeus, flanked by a pantheon of deities, held court, they were overwhelmed with reverence. Their eyes glistened at the sight of the Elysian Fields, where ancient heroes basked in eternal glow, and the gentle ebb of the River Styx, a sanctuary where souls embraced serenity. And in the heart of this celestial tapestry, they discovered a profound truth: the souls of Lahaina had been intertwined with theirs, residing in a realm of unparalleled, peaceful ecstasy. Back on Earth, tales of their journey spread like wildfire. Humanity, infused with renewed vigor, began to appreciate the sanctity of the cosmos. Efforts were redirected from intrusive explorations to harmonious coexistence. Telescopes evolved from mere instruments of observation to vehicles of communication, relaying messages and knowledge between realms.

As decades passed, Earth's climate began to stabilize. The celestial entities, magnanimous in nature, shared ancient wisdom, unveiling the secrets of sustainable living and harnessing natural energies. This new understanding guided the people of Earth in caring for their living, breathing planet. They learned that Earth's natural resources must remain within her to maintain her vitality and strength.

"Only take from her what you must for survival, not what you desire to amass unreasonable wealth for a few." Consequently, the once-rampant natural disasters waned as nature and mankind embraced a harmonious rhythm.

Educational institutions began to incorporate the tales of Elysium into their curricula. Children grew up recognizing the stories of gods and heroes not as myths, but as neighbors in the expansive universe. Temples and observatories, once separate in purpose, fused into centers of spiritual and scientific enlightenment. Yet, the most profound transformation occurred within people's hearts. Wars ceased as nations recognized the futility of terrestrial conflicts when viewed against the backdrop of the cosmos. A new era of peace and cooperation unfurled, guided by the shared wisdom of Earth and Elysium.

The Celestial Wars, once a looming threat, metamorphosed into a beacon of hope. It stood as a testament to the power of understanding, reminding generations that even amid cosmic challenges, unity and knowledge perpetually illuminate the path forward. Chronicles of Unity: The Celestial Concord Over time, the Elysium-Earth Alliance, often dubbed the Celestial Concord, evolved into a symbol of harmony. This newfound unity ushered in advancements unattainable by either realm alone. The gods, once stewards of their sanctified domains, transformed into mentors and allies for humanity.

Athena, goddess of wisdom, established academies where Earth's most brilliant minds collaborated with celestial beings. Together, they deciphered the universe's enigmas, melding divine insights with human inventiveness.

Hephaestus, the god of blacksmithing, imparted cosmic metallurgy secrets, leading to innovations that merged ethereal and tangible realms.

Music and art experienced a renaissance. Apollo and the Muses inspired artists, interweaving earthly melodies with celestial harmonies, crafting symphonies that resonated across dimensions. Painters, swayed by cosmic hues, produced masterpieces that appeared to shimmer with stardust.

Trade and commerce flourished as well. Midgard, Odin's dominion, transformed into a vibrant marketplace where terrestrial goods exchanged hands for celestial artifacts. The once-guarded Bifrost, now a rainbow bridge of vibrant hues, metamorphosed into a bustling trade conduit.

Nevertheless, not all welcomed this unity. A faction emerged, comprised of both human and divine beings, deeming the alliance a dilution of purity. Led by Loki, the cunning trickster god, and joined by those nostalgic for an era when a dominant force reigned over the less fortunate, this faction sought to destabilize the Celestial Concord—such is the nature of malevolent intent.

Loki's mischief knew no bounds. He sowed seeds of doubt, spinning tales of treachery, casting both sides into a wary state. Natural phenomena, once revered and comprehended, now stirred suspicion. Meteor showers and solar flares now bore potential omens of betrayal.

Tensions escalated. The Oracle of Delphi, emblematic of unity, fell victim to vandalism. Temples and observatories encountered targeted acts of destruction. The hard-won peace teetered on the precipice of shattering.

Acknowledging the gravity of the situation, a council convened on neutral ground—the moon, an impartial observer of the evolving Earth-Elysium relationship. Odin and Zeus, flanked by Earth's chosen emissaries, gathered to address the brewing tempest.

In the moon's argent halls, a revelation transpired. Artemis, moon goddess, unveiled visions of a shared past, where gods and mortals coexisted, entwined as harmonious souls. These visions depicted instances of love, sacrifice, and intertwined destinies.

Stirred by these revelations, the council proclaimed a festival—the Celestial Carnival, a celebration of the intertwined destinies of both realms. This event would serve as a commemoration of their shared history and a testament to their united future.

The carnival diverged from convention. Floats bedecked with stars and galaxies graced the thoroughfares. Songs from antiquity, narrating tales of love between gods and mortals, resonated through the atmosphere. Mortals and deities danced beneath the cosmic canopy, their energies merging in a luminous display.

Loki's dissent dissolved amid the deluge of unity. Observing the collective might of two realms, the trickster god withdrew into the shadows, comprehending that division found no place in this emerging epoch.

The Celestial Carnival evolved into an annual tradition, symbolizing the enduring bond uniting Earth and Elysium. It stood as a guiding light for generations, imparting the principles of unity, comprehension, and the enchantment born when disparate worlds merge as one.

The Elysium Accord

In the years that followed the Celestial Carnival, a new understanding began to emerge among Earth's inhabitants.

The rampant exploration of yesteryear, once seen as a testament to human curiosity, was now being re-evaluated. The realization dawned that space, with its vast expanse, was not just a void waiting to be explored, but a territory teeming with life and civilizations that had existed long before humanity looked to the stars.

A new ethos began to permeate scientific communities and global leadership. The United Nations established a Cosmic Diplomacy Division, dedicated to fostering relationships with celestial entities and ensuring Earth's exploratory activities respected the sovereignty of other realms.

A historic event marked this transformation. Earth's leaders, along with representatives from various celestial domains, gathered at the Oracle of Delphi, the sacred nexus between realms. Here, the Elysium Accord was drafted—a comprehensive treaty outlining the rights of exploration, ensuring mutual respect, and establishing protocols for inter-realm interactions.

The key tenet of the Elysium Accord was the principle of "Informed Exploration." Before embarking on any cosmic venture, Earth's explorers would now seek permission, understand the implications of their journey, and ensure their actions did not infringe upon the rights and territories of other celestial entities. As a gesture of goodwill, Odin and Zeus inaugurated an inter-realm portal at the Temple of Concordia.

This portal allowed for designated envoys and explorers to traverse realms, ensuring transparency and fostering trust.

With the Accord in place, Earth's approach to cosmic exploration underwent a profound shift. Instead of dominating and claiming, humans now sought to understand and collaborate. The tales of Elysium, once seen as myths, became subjects of respectful study, bridging the knowledge gap and enriching Earth's own history.

This transformation was not without challenges. There were factions resistant to change, clinging to old ambitions of cosmic conquest. However, the overwhelming consensus among Earth's inhabitants was clear: the universe was a shared tapestry, not a frontier to be seized.

The final chapter of this saga was penned by a joint collaboration between Earth's poets and Elysium's bards. Titled "The Hymn of Harmony," it encapsulated the journey of two realms, from conflict to unity, and the newfound wisdom that exploration, when devoid of hubris, can be a path to enlightenment.

It read:

"In the dance of stars and time,

Two realms once clashed, then aligned.

From discord, unity did birth,

Elysium met humble Earth.

In stars, we sought to stake our claim,

Yet learned respect, not fame.

For in the vast cosmic sea,

Boundaries exist, even if unseen.

Now, with hearts open and hands entwined,

We journey forth, with respect in mind.

For the universe, in all its might,

Is a shared realm of wonder and light."

And so, the people of Earth embarked on a new era, understanding that true exploration was not about conquest but about seeking knowledge with humility and respect. The universe, once a vast unknown, became a testament to the harmonious potential of collaboration and mutual understanding."

CHAPTER TEN

Mysteries of the Cosmos: Insights from the Dimensional Anthropology Museum

Greetings, explorers of the universe! From the hallowed halls of the Dimensional Anthropology Museum, we present a curated collection of the universe's enigmas that have intrigued our esteemed scholars for eons. Embark on this journey of cosmic curiosities with us.

1. Dark Matter: The Celestial Enigma In the vast tapestry of the universe, dark matter acts as its unseen threads. The observable cosmos, from swirling galaxies to vast clusters, finds its structure interwoven with this concealed element. While its gravity reveals its presence, the true essence of dark matter is yet to be grasped. Our Museum's archives brim with theories, from exotic particles to alternative gravitational blueprints.

2. Dark Energy: The Ethereal Expander As dark matter binds, dark energy propels. Constituting a staggering 68% of the universe, it is the force behind the universe's accelerating dance. Despite its dominance, dark energy remains one of the Museum's most sought-after exhibits, its true essence still evading our finest scholars.

3. The Universe's Destiny: Cosmic Prophecies What grand finale awaits our universe? Various cosmic prophecies abound in our Museum. The "Big Crunch" foretells a dramatic gravitational finale, while the "Big Freeze" paints a future of chilling stillness. And then there's the "Big Rip," a cataclysmic tearing apart of everything we know.

4. Black Holes: The Galactic Enchanters These celestial sorcerers have always fascinated our Museum's patrons. Their ethereal presence has been affirmed by recent cosmic observations.

Yet, the mysteries they harbor—from their heart's singular point to the riddles of lost information - remain at the forefront of our studies.

5. The Fermi Paradox: Echoes of Silence The vast cosmic theater, teeming with stars and worlds, seems primed for myriad stories. Yet, the silent vastness perplexes our finest minds. Are we an anomaly, a rarity, or are we just not attuned to the universe's grand opera?

6. Ultra-High-Energy Cosmic Rays: The Astral Artillery These cosmic emissaries, with their boundless energy, are amongst the Museum's prized exhibits. Their origins, whether from the heart of black holes or colossal stellar detonations, continue to enthrall our researchers.

7. The Great Attractor: The Cosmic Magnet: This gravitational enigma beckons entire galaxies, ours included, towards a mysterious rendezvous point. Behind the veil of the Milky Way, the Great Attractor remains one of our Museum's most enticing unsolved puzzles.

8. Fast Radio Bursts (FRBs): The Universe's Whispers: These fleeting cosmic signals, brief yet enigmatic, have set our Museum abuzz since their discovery. Are they cries of distant cosmic entities or echoes of astronomical collisions?

9. Exoplanet Atmospheres: Breaths of Distant Worlds In our Museum's exoplanetary wing, the dream of understanding alien atmospheres comes alive. Every new discovery offers tantalizing hints of worlds both familiar and unimaginably alien.

10. Matter and Antimatter Asymmetry: The Primordial Imbalance, The universe's choice for matter over antimatter, is a question that has graced our Museum's lecture halls for eons. Delving into the dawn of time, our scholars seek answers in both particle collisions and cosmic symphonies.

These brief glimpses represent but a fraction of the cosmic wonders housed within the Dimensional Anthropology Museum. As we continue our quest for knowledge, the universe never ceases to amaze and inspire.

In my heart, following this research and the creation of Homer's final Oddessy, I know that I have amassed great riches, for I find myself devoid of further requests to place before God. The legacy of my journey shall endure as a testament to the harmonious interplay of human ingenuity, divine mysteries, and the unbreakable bond with nature.

For that, let me share with you two additional discoveries. The first one was found within the same James Webb photograph, which comprises a significant portion of the images featured in this publication. The second photograph, transformed into an exquisite poster available for purchase, captures a figure we encountered in "Comus," Maryland, back in 2012.

This photograph seemingly portrays former President Theodore Roosevelt—or more accurately, the left side of Teddy's face. However, preceding our exploration of Mr. Roosevelt, the photograph below holds profound significance. It unveils a facet of early human history that has remained undocumented.

Title: The Bison Man: Unveiling Innovative Prehistoric Cold-Weather Attire

Abstract: This research paper presents an intriguing discovery that sheds light on a creative adaptation by early humans to combat freezing conditions. Through examining the final hit of Iroquois Technology discovered during this research, we unveil an image resembling an individual of Native American or Asian descent.

Utilizing dynamic filtration filters, we enhance the features present in the gas-filled image. Gradually, the visage of The Bison Man's attire emerges. Our researchers posit that the revelation is of a man draped in a dense bison pelt, with particularly intriguing aspects involving the incorporation of the bison's front legs and hooves. Treat this photograph like the others in this book; don't expect immediate clarity upon first glance.

Many of these images require time to manifest. The human eye was never designed for these views; we are granted this glimpse only by divine will. Thus, we invite readers to revisit these pages until the images become clearer.

It appears this individual has taken down a small bison and harnessed the "Hump Hair" as an immense overcoat. The hump of a bison is a blend of muscle, fat, and connective tissue, with the unique hair potentially guarding the bison against harsh elements like snow, rain, and wind. This insulation likely served this native's purpose.

We delve into the potential motives behind this distinct attire, suggesting it fulfilled multifaceted roles —primarily aiding survival and resource transportation.

Introduction: The Bison Man represents a captivating discovery illuminating the resourcefulness and innovative adaptations of early humans surviving in harsh environments. Our focus is unraveling the significance of the Bison Man's attire, characterized by the incorporation of bison skin and the preservation of front legs and hooves.

Methods: Initially inconspicuous, the Bison Man chooses to retain the bison pelt's front legs. This arrangement prompts investigation into the possible motivations and functions behind this unique choice.

Discussion: We propose that the Bison Man's attire served a dual purpose. First, the strategic use of bison skin over shoulders and around the neck offered insulation and protection from frigid temperatures. The sturdy leg bones likely supported and elevated gathered resources. Notably, this attire potentially negated the need for separate hand coverings, as hands could rest within the legs of the bison fur, providing warmth and a secure grip around the bison's lower ankles.

Conclusion: The Bison Man's attire, a blend of innovation and adaptability, provides a glimpse into the ingenuity of early humans surviving harsh climates. While historical records lack explicit mention of this practice, the Bison Man underscores prehistoric individuals' remarkable abilities to devise effective survival strategies. One must speculate whether this individual's fate was sealed by venturing too far from home, encountering an unexpected winter freeze. If one examines where the left bison leg extends along the man's left breastbone, avian feathers and a pointed rostrum appear. Further observation of the bison pelt reveals more furry animals suspended from the bison's legs. If Iroquois Technology indeed captured the figure of an early prehistoric man employing an undocumented technique, this stands as one of the most enthralling discoveries in our research and publication. The initial photograph presents a triptych, featuring three images juxtaposed to display the unaltered depiction of the bison man as discovered through Iroquois technology. The subsequent image portrays the result following the application of dynamic filtration, akin to the transformation carried out on the third photograph.

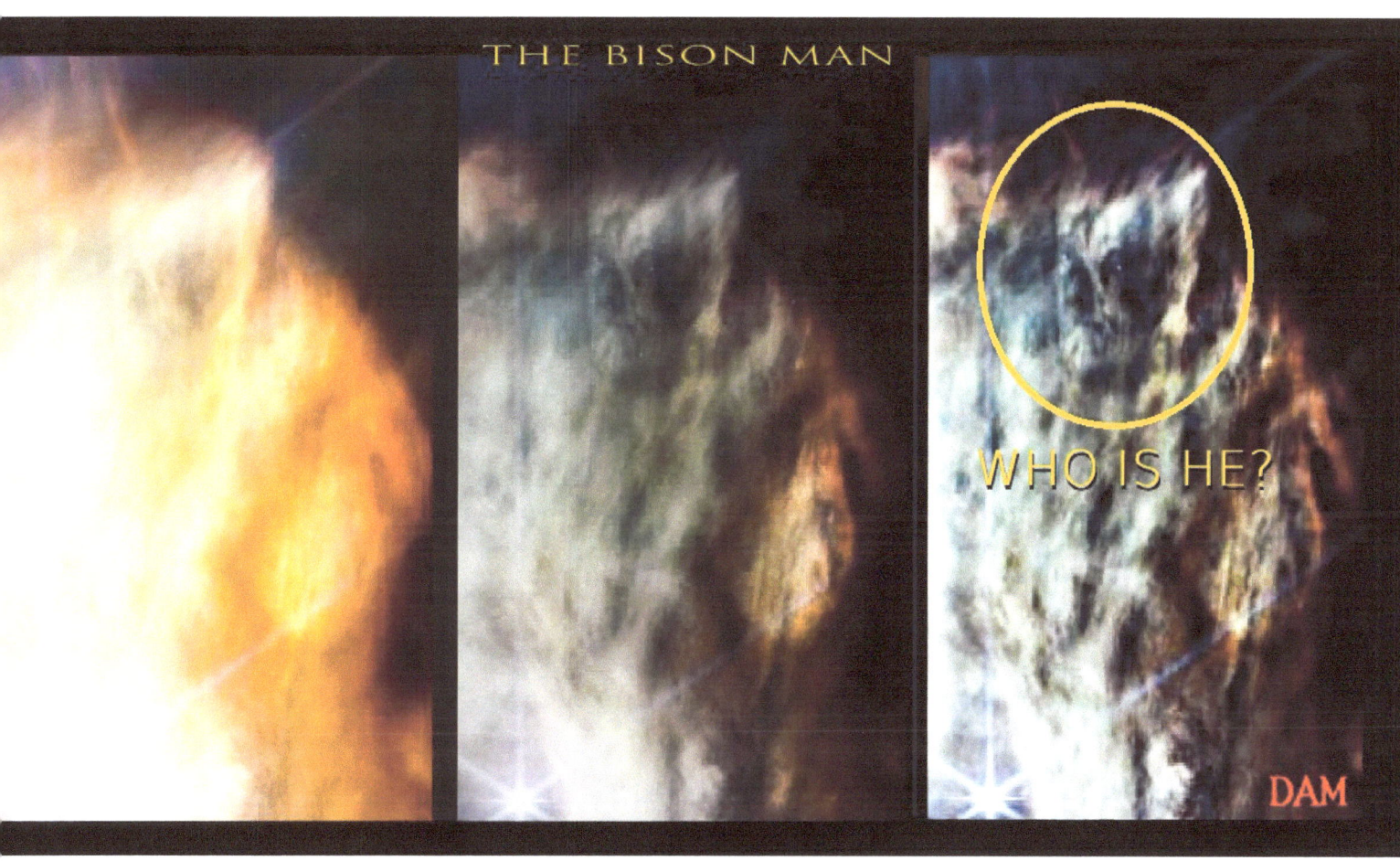

Figure 20: The Bison Man

At the Dimensional Anthropology Museum, researchers have devised a schematic illustration that outlines our interpretation of how the bison man likely donned the bison pelt, with special attention to the utilization of the bison's legs to facilitate lifting and transporting his prey.

Figure 21: Schematic Illustration of TBM

"And last but not least, a photograph that looks remarkably like Theodore "Teddy" Roosevelt, who was America's 26th U.S. president from 1901 to 1909, is celebrated for his vibrant spirit, environmental initiatives, and pivotal role in the inception of the National Parks system.

Figure 22: The half-face of Roosevelt

As a leading figure in the Progressive Era, he fervently pushed for reforms, confronting corporate monopolies and advocating for the "Square Deal" for Americans. Notably, Roosevelt was a Rough Rider in the Spanish-American War and was posthumously recognized with the Medal of Honor for his bravery. Accompanying this article is a juxtaposed photograph: on the left is a familiar image of Roosevelt, while the right showcases a mysterious figure identified by Iroquois Technology in Comus, Maryland in 2012. Although the Dimensional Anthropology Museum has not confirmed whether this image, dubbed QEIN, truly represents Teddy, its captivating nature is undeniable. Our team's belief that the figure could be Roosevelt is bolstered by two separate sightings in the same area. In a subsequent capture, images suggestive of his sons, Archibald and Teddy Jr., added to the intrigue. These remarkable photographs, along with about 100 other astounding discoveries from almost twenty years of research, will be featured in the upcoming exhibit at the Dimensional Anthropology Museum, including some that intriguingly resemble dinosaurs.

The reason we cannot display the entire face of the image that looks like Mr. Roosevelt is that only one side of the face could be seen in the original video from where the figure peered around the side of a large tree. It was as if he only wanted the left side of his face to be seen. In another image, we captured years apart from this image. We did see the right side of his face. But we were confused as it clearly looked like his right eye was damaged in some way. We were convinced at this point it was not the captured figure of Teddy Roosevelt as it was not widely known that he had a bad right eye. However, throughout research, we later discovered that Theodore Roosevelt suffered a detached retina in his right eye due to a boxing injury that he sustained while he was in office. In 1908, Roosevelt was sparring with a young military aide, Captain Archibald Butt, in a boxing match at the White House. During the match, a blow from Captain Butt's fist struck Roosevelt's right eye, causing a detached retina.

The injury led to vision impairment in his right eye, and despite medical treatment, his vision never fully recovered. The incident and resulting injury had a lasting impact on Roosevelt's eyesight and health. The vision impairment did not deter him from continuing his political career and public activities, but it was a reminder of the physical toll his active lifestyle could take.

Roosevelt's detached retina is a notable part of his medical history, but very few Americans today know about this fact. Allow me this moment to urge you to immerse yourself in the pages of this publication, to embrace the journey it unfolds. In a world where the existence of an afterlife remains a realm of uncertainty—neither proven nor claimed by anyone throughout history—those of us at the Dimensional Anthropology Museum, including myself, remain humbly aware of this fact. It is an essential perspective to bear in mind as we explore the enigmatic realms within these pages. Our intent lies not in the realm of scientific certainty but in the realm of the enthralling and the wondrous.

We brandish the label of entertainment, a testament to the recognition that we tread the line between fascination and empirical certainty.

Amidst these pages, none of the images you encounter are products of human artistry; they are manifestations of nature herself. Nature, with her veils of mystery, harbors enigmas that may forever elude our understanding. This compendium of revelations merely contributes to the ever-expanding tapestry of those enigmas. It is paramount to acknowledge that the purpose of this book is to entertain and enchant, not to assert dogmas.

Whether your beliefs find solace in the embrace of a higher power or revel in the realm of skepticism, the contents within these pages should not disrupt nor reshape those convictions. Let not this book serve as a tool of persuasion; its purpose is to provoke, to inspire contemplation. This tome and the museum stand as beacons that beckon those who are curious to envision the universe through unbounded lenses, challenging the constraints of preconceived notions.

Within these images, we discover the liberty of autonomous thought—a gift bestowed upon us by a higher realm. It is disheartening to observe a society where conformity often overshadows individual ruminations. Reclaim your prerogative to shape your universe according to your ideals. You are under no obligation to share these musings; they are your intimate tapestries, as cherished as newborn hopes.

Dear Esteemed Researchers, Dedicated Fans, and Selfless Volunteers, for over two decades, you have journeyed with us through the vast realms of dimensional anthropology, aiding us in our mission to uncover the complexities of our universe. From our very first exhibition to our most recent discoveries, each one of you has played an indispensable role in crafting the legacy of the Dimensional Anthropology Museum.

First and foremost, to our researchers – your tireless pursuit of knowledge, your ceaseless curiosity, and your unwavering dedication have propelled us to the forefront of dimensional research. The groundbreaking work you have spearheaded, the evidence revealed, and the digital artifacts collected have opened the minds of countless individuals to the wonders of the cosmos. Our institution would not stand where it does today without your invaluable contributions.

To our fans – you have breathed life into our endeavors. Your passion, enthusiasm, and insatiable quest for understanding have echoed through our research, reverberating the importance of our mission. Your continued patronage and support have not only sustained our research but have also affirmed the significance of the work we undertake.

And to our incredible volunteers – the very heart and soul of this institution. You have dedicated countless hours or field research, offering your skills, time, and energy selflessly.

It is with a heavy heart and profound gratitude that I announce the conclusion of this chapter of our journey together. This book the *Webb Telescope's Revelations: Is Humanity Ready for the Truth?*, filled with wonder, discovery, and exploration, will always remain a testament to what we can achieve when we come together in the name of science.

However, as we stand on the precipice of a changing world, facing challenges unprecedented in scale and impact, our priorities too must evolve. The harrowing realities of global warming, coupled with devastating disasters that have rendered millions homeless and destitute, beckon us to a new calling. It is a call to solidarity, to compassion, and to urgent action.

We find ourselves at a crossroads where we must channel our resources, expertise, and collective efforts towards Disaster Relief. Our commitment to science and understanding, while unwavering, must now be directed towards alleviating the human suffering that these disasters inflict. The time has come for us to shift our focus, to step up and stand beside those most in need.

This is not a departure from our mission but rather an evolution. Science, at its core, is an endeavor to better humanity, and now, more than ever, our fellow human beings need us. As we embark on this new journey, we hope you will join us, bringing to it the same fervor, dedication, and heart that has defined our last two decades.

In closing, thank you, from the deepest recesses of our hearts. Your contributions to the Dimensional Anthropology Museum have not only furthered the realms of scientific discovery but have also enriched countless lives. As we transition into our new focus on Disaster Relief, we carry forward the ethos of community, collaboration, and a shared commitment to the betterment of humankind. With deepest gratitude and hope for the journey ahead, TD McRoy your friend in time.

Call to Action:

The future trajectory for the Dimensional Anthropology Museum is primarily centered on disaster relief. As a recognized 501(c)(3) institution, our funding sources are multifarious.

We cordially invite you to explore our website and contribute generously, assisting us in our mission to serve others.

For those eager to delve deeper into the myriad discoveries made by Iroquois Technology over the past two decades, we warmly welcome you to peruse our exclusive Etsy store dedicated to Iroquois Technology. Within its digital confines, you will encounter a rich collection of photographs, taken straight from this book, masterfully transformed into captivating posters. Every purchase you make not only fuels our museum's growth but also significantly aids The Bertha Project's disaster relief initiatives. As you turn the page, be prepared for an enlarged rendition of 'The Bison Man' image. In its intricate detail, discerning eyes might spot a young girl nestled below his right chest, her gaze fixed on the time-worn figure above.

Figure 23: The Bison Man

"What if hurricanes aren't weather events?" The final contribution from Iroquois Technology and Dynamic Filtration. The date was August 29, 2023, and the time was around 22:30 hours. There were mostly weather-related news reports in the Orlando area, as Hurricane Idalia was expected to make landfall on the Northwest Coast of Florida. Since the Iroquois Technology programs were going to be destroyed soon, the program was run on radar footage from the National Weather Channel. The results of the program were fascinating, and the posters are the outcome of the findings.

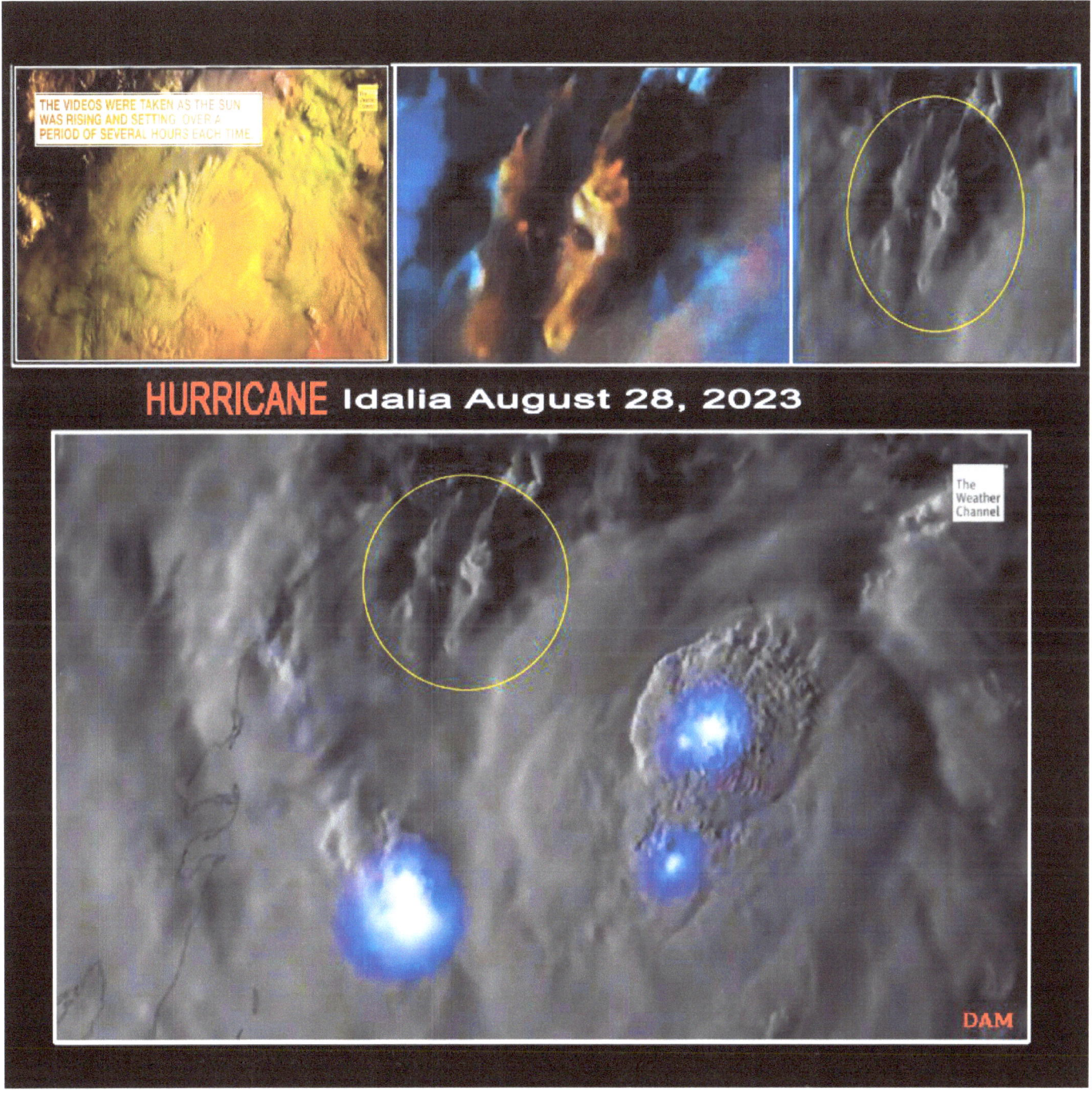

Figure 24: Seminole's vs Muscogee's

Our researchers had little time to evaluate the figures in the weather channel video clip, but two native Indians appeared in the footage. The one moving from the south heading north looks like a Seminole Indian, while the one north of him looks like he could be from the Muscogee (Creek) Indian tribe. To the left of the Muscogee figure, there appears to be a horse with one braid in view from its mane.

Florida's history talks about conflicts between the Seminole Indians and the Muscogee (Creek) Indians. These conflicts were a part of the broader struggles among Native American tribes and European settlers in the southeastern United States during the 18th and 19th centuries. The disputes over land and resources led to tensions and clashes between various tribes.

One of the most significant conflicts involving the Seminole and Muscogee was the Creek War of 1813-1814. The Seminole Wars, a series of conflicts between the Seminole people and the United States, began in the early 19th century and continued into the mid-1800s. These wars were driven by a combination of factors, including disputes over land, cultural differences, and conflicts with American expansionism.

It's important to note that these historical events were complex and involved multiple parties with differing motivations and allegiances. The Dimensional Anthropology Museum researchers were on a tight schedule for this portion to be ready for publication, but we surmise that if it is possible that any of the Greek Mythology discovered in this book is ever deemed to be verified hundreds of years from now. Could this mean that the battles between the Seminoles and the Muscogee tribes are still being waged like they have been since man first stood on land in Florida?

The Dimensional Anthropology Museum was created to blend science with entertainment to offer another perspective on our planet and the cosmos, as we all might be living in a supernatural world. And if so could it be possible that the powerful forces of nature could be easily explained someday? Only time will tell. Please visit our Iroquois Technology store on Etsy to view a large poster of the Hurricane photographs. https://www.etsy.com/shop/IroquoisTechnology

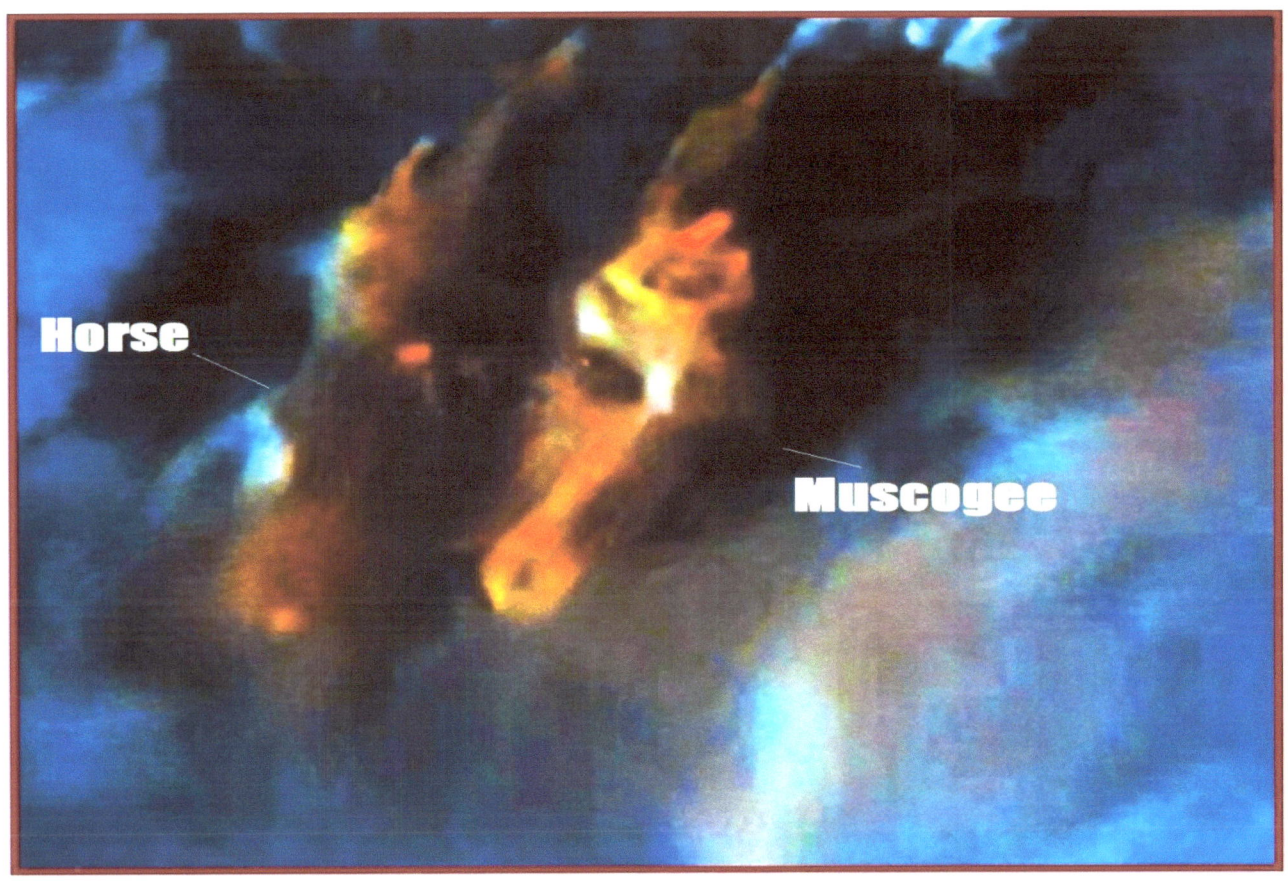

Figure 25: Seminole vs Muscogee

Do the images on the prior page, taken from the National Weather Channel's radar, support the local Floridian belief that Seminole spirits protect Tampa Bay from direct hurricane strikes? While these images could be compelling for casual discussions, such as in a sports bar, one might wonder if the combined presence of both the Muscogee and Seminole references implies a united effort to shield Florida's inhabitants from the hurricane's impact.

1. "On the next page, you will see a collection of figures that Iroquois Technology detected and brought into view using Dynamic Filtration. It is important to note that we cannot confirm the existence of any other species of human on planet Earth, other than what has been taught in schools and universities. However, the computer programs have a big flaw; they can only pinpoint the figures, but not tell us what we are looking at. It is up to the viewer's imagination to interpret the images.

2. "For those who have made it this far into the book, I would like to say thank you. The Dimensional Anthropology Museum's research was never intended for mass publication. Our theories were well outside the boundaries of traditional scientific research, and the conclusions were unfathomable for most. As a result, discovering the images without the pressure to prove our finds was a relaxing hobby.

3. "As humans, we often rush to offer our special talents to the world. But in this case, we never wished to do so. As a research organization, we understand that sharing too much information can lead to the destruction of what we love about what we are doing. Therefore, we limited the information we shared whenever we used the Iroquois Technology in film and TV appearances.

We were sure never to reveal what the program had really captured.

4. "However, now that both programs have been decommissioned, we have decided to share 12 of our most amazing early man figures in poster format. Displayed on the poster, which can be purchased at our Etsy store, "Iroquois Technology Store," are 11 human-like forms and what appears to be a dinosaur lying down beside a tree in Pennsylvania. Upon close observation, it appears to have a severe injury on its left upper flank. We deliberately left out a photograph of another dinosaur that was carrying a saddle on its back with large rigged stirrups, as it would become too confrontational and pull readers away from the true nature of this publication. "Allow me a brief detour from the script. I couldn't help but chuckle when pondering the undoubtedly provocative nature of the images in this publication. Yet, the image suggesting a dinosaur adorned with what seems to be a thick leather saddle was the one I left out. The scientific consensus has long held firm to the belief that humans and dinosaurs never walked the Earth simultaneously. Yet, if any of the figures unearthed in our NASA findings hold water, it would radically overturn this notion. Not only would it suggest that humans and dinosaurs coexisted, but it might also imply that humans harnessed these colossal creatures for transportation. Imagine a world where dinosaurs weren't just neighbors, but also our modes of travel."It would certainly redefine the saying, 'Whoa there, big fella."

5. "The figure that emerged on the poster, discovered in Egypt, was unveiled from within a wall of one of the pyramids during a segment hosted by NBC News Chief Correspondent Richard Engel. Another figure is visible above his head, believed to represent Tutankhamun (formerly known as Tutankhaten), the third-last pharaoh of the Eighteenth Dynasty of ancient Egypt. We aim to raise $250,000 to establish an online immersive museum, allowing the public to view 100 images that have yet to be shared."

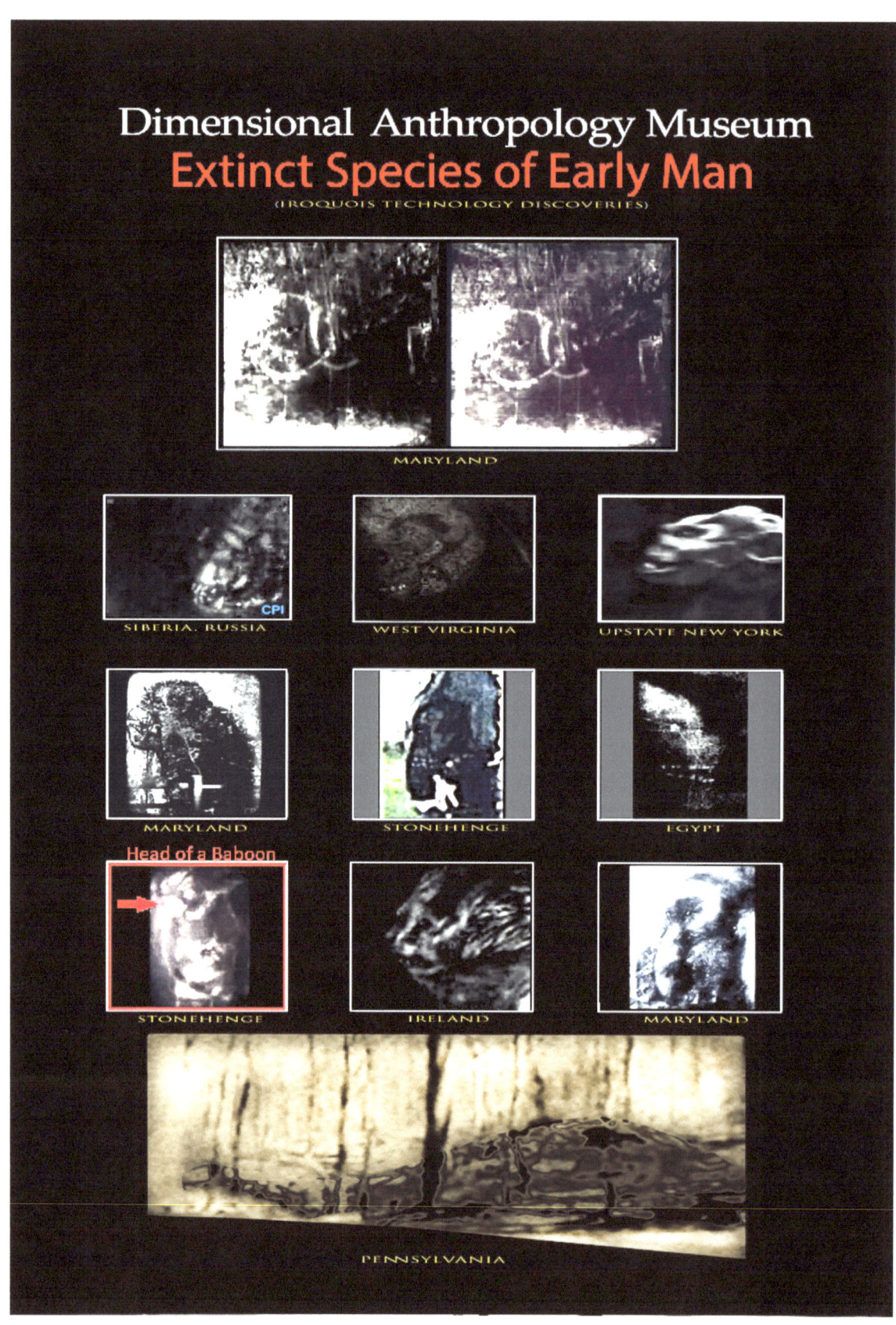

Figure 26: Extinct Species of Early Man

1. "**Evolution of Early Man and The Enigmatic Species of Yore**"

2. "**Introduction:**"

The story of man's evolution has always captivated the human imagination. But as our understanding deepens, we are constantly reminded of how little we truly know.

There have always been theories about the existence of prehistoric human species, and some of these theories have gone as far as suggesting that other species existed billions of years ago, leaving no remnants. In this document, we explore these mysterious lost species and the astounding findings of the Dimensional Anthropology Museum.

3. "**The Evolutionary Timeline:**"

As understood today, our direct ancestors, the Homo sapiens, evolved around 300,000 years ago. Before them, we had Homo erectus, and a lineage of predecessors dating back to about 2 million years ago. But what if our lineage is far older, stretching back billions of years?

4. "**Lost Species – A Possibility?**"

The idea that there were species of man billions of years ago is tantalizing, but it challenges our current understanding of evolution. These beings, if they existed, might have been so different that we cannot even begin to fathom their physiology or way of life. Maybe they were not even carbon-based, or perhaps Earth was a different place altogether.

5. "**The Dimensional Anthropology Museum's Findings:**"

In this publication, the Dimensional Anthropology Museum is sharing images of other human species predating Homo sapiens and Homo erectus. While many should be skeptical, the intricacies and consistencies between the figures and what is seen on the poster cannot be ignored.

6. "**A Tale of Teeth:**"

One of the most striking features in the figures is the diversity of teeth among these early humans. Some show long, sharp canines hinting at a carnivorous diet or perhaps a mechanism for defense or courtship. In contrast, others have flat and thick molars, indicating a diet rich in hard and fibrous vegetation.

7. "**The Fashion Sense of Prehistoric Man:**"

Two startling discoveries emerged from the figures. Firstly, an early man found in what is now modern-day Russia was seen wearing headgear strikingly similar to another early human located in Maryland, USA. Despite being thousands of miles apart, the similarity in design, materials, and construction hints at a shared culture or perhaps a common origin.

In a startling revelation, we unearthed remains of an early human in Egypt, garbed remarkably like a pharaoh. The attire echoes that of Ramesses the Great as immortalized in "The Ten Commandments". Yet, where do we draw the line between art mirroring life and life reflecting art? Now, brace yourself: this pharaoh's countenance is not that of a man but of a baboon.

Picture this: a baboon cloaked in the splendor of pharaonic vestments. Pause for a moment to let that sink in. Could it be conceivable that the pyramids, ancient beyond our common understanding, were erected by an earlier form of mankind? A species more akin to the depictions discovered by Iroquois Technology and showcased in our 'Early Man' poster? Were they the true architects of the marvels we witness today?

8. "**Conclusions and Further Questions:**" The revelations from the Dimensional Anthropology Museum have surely shaken our understanding of early man. While many questions arise, the possibility of diverse early human species and a shared global culture cannot be outrightly dismissed.

Could it be that humanity has risen and fallen many times over billions of years? Or are these just coincidences, fabrications, or misunderstandings of the technology?

While it's tempting to draw immediate conclusions, the prudent path would be thorough research, validation, and open-minded exploration. Only time and science will truly tell the story of our ancient ancestors.

9. "**Postscript:**"

As we continue to unearth our past, one thing is certain: humanity's story is ever-evolving, and every new discovery only adds another layer of depth and intrigue to our shared history. Although the programs were never validated and none of the figures in this publication can be authenticated, the art of imagination and wonder cannot be drained away from who we are. The truth is out there, and God willing, we will know it someday, but for now, we can consider ourselves lucky that we might be part of something grander than science could ever imagine.

Support our mission by donating, and help us showcase more stunning images in our museum. Your kind contribution to The Bertha Project also makes a broader impact: it aids in providing water and essential supplies to those affected by natural disasters and strengthens our PIO force of retired law enforcement officers, which plays a pivotal role in reducing looting in communities.

Dimensional Anthropology Museum

"Webb Telescope's Revelations: Is Humanity Ready for the Truth?
Poster Collection available on Etsy: Iroquois Technology Store"

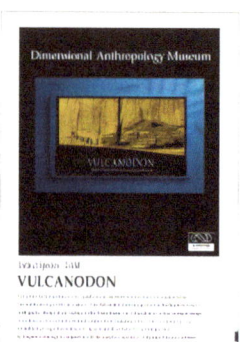

24x36 Brown Vulcanodon

24x36 Early Man Poster

24x36 Odin Poster We Live

24x36 Rainbow Vulcanodon

24x36 Teddy Poster

24x36 The Bison Man Poster

📖 Dive into our curated collection on Etsy: https://www.etsy.com/shop/IroquoisTechnology
🤝 Lend a hand and make a meaningful impact: www.theberthaproject.org
👕 Explore our exclusive retired gear: https://tinyurl.com/bdvn53bd
👕 Browse The Bertha Project's official gear: https://www.theberthaproject.org/theberthaprojectgear

" I know I have achieved great riches, as I have no more favors to ask of God." quote by TD McRoy

Yours sincerely,

TD McRoy